Toward 6G: A New Era of Convergence

Toward 6G: A New Era of Convergence

Amin Ebrahimzadeh

Martin Maier

IEEE
COMMUNICATIONS
SOCIETY

The ComSoc Guides to Communications Technologies
Nim K. Cheung, *Series Editor*
Richard Lau, *Associate Series Editor*

IEEE PRESS
WILEY

Published by John Wiley & Sons, Inc., Hoboken, New Jersey.
Published simultaneously in Canada.

For general information on our other products and services or for technical support, please contact our Customer Care Department within the United States at (800) 762-2974, outside the United States at (317) 572-3993 or fax (317) 572-4002.

Wiley also publishes its books in a variety of electronic formats. Some content that appears in print may not be available in electronic formats. For more information about Wiley products, visit our web site at www.wiley.com.

Library of Congress Cataloging-in-Publication Data

Names: Ebrahimzadeh, Amin, author. | Maier, Martin, 1969- author.
Title: Toward 6G : a new era of convergence / Amin Ebrahimzadeh, Martin
    Maier.
Description: Hoboken, New Jersey : John Wiley & Sons, Inc., [2021] |
    Includes bibliographical references and index.
Identifiers: LCCN 2020034076 (print) | LCCN 2020034077 (ebook) | ISBN
    9781119658023 (paperback) | ISBN 9781119658030 (adobe pdf) | ISBN
    9781119658047 (epub)
Subjects: LCSH: Wireless communication systems–Technological innovations.
    | Network performance (Telecommunication)
Classification: LCC TK5103.2 .E34 2021 (print) | LCC TK5103.2 (ebook) |
    DDC 621.3845/6–dc23
LC record available at https://lccn.loc.gov/2020034076
LC ebook record available at https://lccn.loc.gov/2020034077

Cover Design: Wiley
Cover Image: © John Wiley & Sons, Inc

10 9 8 7 6 5 4 3 2 1

For my soulmate, Atefeh, who dreams and who knows magic is real.
— Amin Ebrahimzadeh
To Alexie and our two children Coby and Ashanti Diva. May J. M. Keynes'
"Economic Possibilities" predicted for 2030 become a reality for them.
— Martin Maier

Contents

Author Biographies

Amin Ebrahimzadeh received the BSc[S3G1] and MSc degrees in Electrical Engineering from the University of Tabriz, Iran, in 2009 and 2011, respectively, and the PhD degree (Hons.) in telecommunications from the Institut National de la Recherche Scientifique (INRS), Montréal, QC, Canada, in 2019. From 2011 to 2015, he was with the Sahand University of Technology, Tabriz, Iran. He is currently a Horizon Post-Doctoral Fellow with Concordia University, Montréal. His research interests include Tactile Internet, 6G, FiWi networks, multi-access edge computing, and multi-robot task allocation. He was a recipient of the doctoral research scholarship from the B2X program of Fonds de Recherche du Québec-Nature et Technologies (FRQNT).

Martin Maier is a full professor with the Institut National de la Recherche Scientifique (INRS), Montréal, Canada. He was educated at the Technical University of Berlin, Germany, and received MSc and PhD degrees both with distinctions (summa cum laude) in 1998 and 2003, respectively. He was a recipient of the two-year Deutsche Telekom doctoral scholarship from

1999 through 2001. He was a visiting researcher at the University of Southern California (USC), Los Angeles, CA, in 1998 and Arizona State University (ASU), Tempe, AZ, in 2001. In 2003, he was a postdoc fellow at the Massachusetts Institute of Technology (MIT), Cambridge, MA. Before joining INRS, Dr. Maier was a research associate at CTTC, Barcelona, Spain, 2003 through 2005. He was a visiting professor at Stanford University, Stanford, CA, 2006 through 2007. He was a co-recipient of the 2009 IEEE Communications Society Best Tutorial Paper Award. Further, he was a Marie Curie IIF Fellow of the European Commission from 2014 through 2015. In 2017, he received the Friedrich Wilhelm Bessel Research Award from the Alexander von Humboldt (AvH) Foundation in recognition of his accomplishments in research on FiWi-enhanced mobile networks. In 2017, he was named one of the three most promising scientists in the category "Contribution to a better society" of the Marie Skłodowska-Curie Actions (MSCA) 2017 Prize Award of the European Commission. In 2019/2020, he held a UC3M-Banco de Santander Excellence Chair at Universidad Carlos III de Madrid (UC3M), Madrid, Spain.

Foreword

A new generation of cellular standards was introduced by the industry once every 10 years since 1979. Each generation provides a big improvement in performance, functionality, and efficiency over the previous generation. These standards were driven mainly by the International Telecommunication Union Radio Communication Sector (ITU-R) and the third generation partnership project (3GPP). As 5G started deployment in 2019, different study groups are poised to examine the possibility of 6G to appear around 2030. One such study group is the ITU-T Focus Group on Technologies for Network 2030. In May 2019, the group issued a white paper entitled "Network 2030 – A Blueprint of Technology, Application and Market Drivers Towards the Year 2030 and Beyond." Among the new applications being studied by the group are holographic media and multi-sense communication services which include transmission of touch and feel as well as smell and taste, in addition to sight and sound that we already enjoy today. Such new applications are expected to give rise to a brand new class of vertical market in entertainment, healthcare, automotive, education, and manufacturing.

It is perfect timing for researchers Amin Ebrahimzadeh and Martin Maier to write their book on "Toward 6G: A New Era of Convergence." The authors surveyed the literature on different 6G proposals including their own work and wrote this book on what 6G would look like in the future. 6G is expected to be built on the strong foundation of 5G, in particular its ultra-high speed and reliability with ultra-low latency. These features enable 6G to support new applications involving human senses such as haptic communication as in the Tactile Internet, as well as high-resolution immersive media beyond today's virtual reality (VR) and augmented reality (AR). The transmission of realistic hologram involves sending volumetric data from multiple viewpoints to account for the 6 degrees of freedom (tilt, angle, and shift of the observer relative to the hologram). The authors provided quantitative examples of such 6G applications requiring the complex interplay of human, robots, avatars, and sophisticated digital twins of objects.

I am particularly intrigued by the last chapter, where the authors summarized their discussions in earlier chapters as the evolution to the "Internet of No Things" in the 6G post-smartphone era, in which smartphones may not be needed anymore. They presented the concept of extended reality (XR) which spans the continuum from pure reality (offline) at one end to pure virtuality (online) at the other end. The middle of the continuum is the region of mixed reality that covers the space from AR to Augmented Virtuality. The authors further expanded the XR concept to extrasensory perception (ESP) as a nonlocal awareness of space and time, mimicking the principle of nonlocality of the quantum realm. The authors undoubtedly provided us plenty of food for thought as we continue our journey from the well-defined 5G standards to the new world of 6G.

Nim Cheung

26 May 2020

Preface

In March 2019, I was approached to publish a book with Wiley-IEEE Press to give visibility to our pioneering work on fiber wireless access. After a short period of reflection, I was willing to accept the invitation and prepare a manuscript, making the following two suggestions. First, we should extend the scope of the book significantly by including technologies that are starting to play a key role in the future 6G vision. Based on the position taken in a commissioned paper back in 2014, where I advocated that we enter an age of convergence, I suggested that 6G will not be a mere exploration of more spectrum at high-frequency bands, but it will rather be a convergence of upcoming technological trends, most notably connected robotics, extended reality, and blockchain technologies. Second, I suggested to involve Dr. Amin Ebrahimzadeh as lead author, with whom I have been closely collaborating on those research topics during his doctoral and postdoctoral studies over the last four to five years, while my role will be more that of a spiritus rector, much like a quarterback in modern American football. Gratefully, our Wiley-IEEE book proposal was very well received by all reviewers and the book project was underway to become the first book on 6G.

What will 6G be? Among others, 6G envisions four-tier network architectures that will extend the 5G space-air-ground networks by integrating underwater networks and incorporating key enabling technologies such as millimeter-wave and Terahertz communications as well as brand-new wireless communication technologies, most notably reconfigurable intelligent surfaces. Furthermore, 6G will take network softwarization to a new level, namely toward network intelligentization. Arguably more interesting, while smartphones were central to 4G and 5G, there has been an increase in wearable devices (e.g., Google and Levi's smart jacket or Amazon's recently launched voice-controlled Echo Loop ring, glasses, and earbuds) whose functionalities are gradually replacing those of smartphones. The complementary emergence of new human-centric and human-intended Internet services, which appear from the surrounding environment when needed and disappear when not needed, may bring an end to smartphones and potentially drive a majority of 6G use cases in an anticipated post-smartphone era. Given that the smartphone is sometimes called the new cigarette of the twenty-first century and using it is considered the new smoking, the anticipated 6G post-smartphone era may allow us to rediscover the offline world by co-creating technology together with a philosophy of technology use toward *Digital Minimalism*, as recently suggested by computer scientist Cal Newport.

As this book is ready to go to press, the currently most intriguing 6G vision out there at the time of writing was outlined by Harish Viswanathan and Preben E. Mogensen, two Nokia Bell Labs Fellows, in an open access article titled "Communications in the 6G Era" that was published just recently last month. In this article, the authors focus not only on the technologies but they also expect the human transformation in the 6G era through unifying experiences across the physical, biological, and digital worlds in what they refer to as *the network with the sixth sense*. This book aims at providing a comprehensive overview of these and other aforementioned developments as well as up-to-date achievements, results, and trends in the research on next-generation 6G mobile networks.

Martin Maier

Montréal, April 2020

Acknowledgments

The completion of this book would have never been possible without the support and collaboration of a number of amazing people. We would like to thank Professor Eckehard Steinbach, Dr. Claudio Pacchierotti, and Dr. Leonardo Meli for providing us with the teleoperation and telesurgery traces. We thank Abdeljalil Beniiche for his collaboration in surveying the state-of-the-art of blockchain technologies and developing our proposed nudge contract in Chapter 6. Special thanks go to Sajjad Rostami for his endless efforts in our lab toward developing the experimental framework used in Chapter 7. In particular, we are grateful to Nim Cheung, the former President of IEEE Communications Society, who invited Martin to write this book, as a new entry to the ComSoc Guide to Communications Series. At Wiley-IEEE Press, we would like to thank Mary Hatcher, Victoria Bradshaw, Louis Vasanth Manoharan, and Teresa Netzler for their guidance throughout the whole process of preparing the book. We would like to acknowledge the Natural Sciences and Engineering Research Council of Canada (NSERC) and the Fonds de Recherche du Québec-Nature et Technologies (FRQNT) for funding our research. Finally, and most importantly, Amin would like to take this opportunity to express his great depth of gratitude to his parents for their endless support, love, and encouragement.

Acronyms

1G	First generation
2G	Second generation
3G	Third generation
3GPP	3rd generation partnership project
6Genesis	6G enabled smart society and ecosystem
6GFP	6Genesis flagship program
A2A	Avatar-to-avatar
A2H	Avatar-to-human
ACCs	Access control contracts
ADC	Analog-to-digital converter
AGI	Artificial general intelligence
AI	Artificial intelligence
ANN	Artificial neural network
API	Application programming interface
APT	Advanced persistent threat
AR	Augmented reality
ART	Audi robotic telepresence
AV	Augmented virtuality
B5G	Beyond 5G
BBU	Baseband unit
BIoT	Blockchain-based IoT
BS	Base station
CAeC	Contextually agile eMBB communications
CAPSTA	Context-aware prioritized scheduling and task assignment
CCDF	Complementary cumulative distribution function
CCSC	Crypto currency smart card
CNRS	Centre National de la Recherche Scientifique

CoC	Computation oriented communications
co-DBA	Cooperative dynamic bandwidth allocation
CoMP	Coordinated multipoint
CPRI	Common public radio interface
CPU	Central processing unit
C-RAN	Cloud radio access network
DAC	Digital-to-analog converters
DAO	Decentralized autonomous organization
DApps	Decentralized applications
DBA	Dynamic bandwidth allocation
DC	Direct current
DCF	Distributed coordination function
DFR	Decreasing failure rate
DIFS	DCF interframe space
DLT	Distributed ledger technology
DNS	Domain name system
DoF	Degrees-of-freedom
DSOC	Decentralized self-organizing cooperative
DVB	Digital video broadcasting
DVS	Dynamic voltage scaling
ECDSA	Elliptic curve digital signature algorithm
eMBB	Enhanced mobile broadband
EPON	Ethernet passive optical network
ESF	Edge sample forecast
ESPN	Extrasensory perception network
EVM	Ethereum virtual machine
FiWi	Fiber-wireless
FRF	Failure rate function
FTTN	Fiber-to-the-node
FTTx	Fiber-to-the-x
Fx-FH	Fx fronthaul
GP	Generalized Pareto
GSM	Global system for mobile communication
HABA	Humans-are-better-at
HART	Human-agent-robot teamwork
HITL	Human-in-the-loop
HMI	Human–machine interaction
HSI	Human system interface
I2V	Invisible-to-visible
IA	Intelligence amplification
ICT	Information and communication technology

IFR	Increasing failure rate
IMT 2020	ITU's international mobile telecommunications 2020
IoE	Internet of everything
IoS	Internet of skills
IoT	Internet of Things
IP	Internet protocol
IPACT	Interleaved polling with adaptive cycle time
ITU-T	ITU's telecommunication standardization sector
JC	Judge contract
JND	Just noticeable difference
KPI	Key performance indicator
LoRa	Long range
LPWA	Low-power wide-area
LTE	Long-term evolution
LTE-A	LTE-advanced
M2M	Machine-to-machine
MABA	Machines-are-better-at
MAC	Medium access control
MAP	Mesh access point
MCC	Mobile cloud computing
MEC	Multi-access edge computing
MIMO	Multiple-input multiple-output
MLE	Maximum likelihood estimation
MLP	Multi-layer perceptron
mMTC	Massive machine type communications
mmWave	Millimeter-wave
MP	Mesh point
MPCP	Multipoint control protocol
MPP	Mesh portal point
MR	Mobile robot
MU	Mobile user
NAT	Network address translation
NG-PON	Next-generation PON
NOMA	Non-orthogonal multiple access
OFDM	Orthogonal frequency division multiplexing
OLT	Optical line terminal
ONU	Optical network unit
OPEX	Operational expenditures
PDF	Probability distribution function
pHRI	Physical human–robot interaction
PON	Passive optical network

PoS	Proof-of-stake
PoW	Proof-of-work
QoE	Quality of experience
QoS	Quality of service
qubit	Quantum bit
R&F	Radio-and-fiber
RACS	Remote APDU call secure
RF	Radio frequency
RIS	Reconfigurable intelligent surface
RoF	Radio-over-fiber
RRH	Remote radio head
RTP	Real-time transport protocol
SDN	Software-defined networking
SDONs	Software-defined optical networks
SDR	Software-defined radio
SDS	Software-defined surface
SLAM	Simultaneous localization and mapping
SMS	Short message service
STA	Station
TDM	Time division multiplexing
THz	Terahertz
TLD	Top-level domain
ToD	Teleoperated driving
TOR	Teleoperator robot
UAV	Unmanned aerial vehicle
UDP	User datagram protocol
URLLC	Ultra-reliable and low-latency communications
UX	User experience
VHT	Very high throughput
VR	Virtual reality
WDM	Wavelength division multiplexing
WLAN	Wireless local area network
XR	Extended reality

1

The 6G Vision

1.1 Introduction

With the completion of third generation partnership project (3GPP) Release 15 of the 5G standard in June 2018, the research community has begun to shift their focus to 6G. In July 2018, ITU's Telecommunication standardization sector (ITU-T) Study Group 13 has established the *ITU-T Focus Group Technologies for Network 2030 (FG NET-2030)*. FG NET-2030 will study the requirements of networks for the year 2030 and beyond and will investigate future network infrastructures, use cases, and capabilities. According to Yastrebova et al. (2018), current networks are not able to guarantee new application delivery constraints. The application time delivery constraints will differ in terms of required quality of service (QoS). For instance, for Internet of things (IoT) applications, the delay can be up to 25 ms, but connected cars will need 5–10 ms to get information about road conditions from the cloud to make the drive safe. Current cellular networks are not able to guarantee these new application delivery constraints. For illustration of these shortcomings, the authors of Yastrebova et al. (2018) mentioned that the end-to-end latency in today's 4G long-term evolution (LTE) networks increases with the distance, e.g. 39 ms are needed to reach the gateway to the Internet and additional 5 ms are needed to receive a reply from the server. Furthermore, the number of active devices per cell greatly affect the network latency. Measurements of highly loaded cells showed an increase of the average latency from 50 to 85 ms. Among others, the authors of Yastrebova et al. (2018) expect that future mobile networks will enable the following applications:

- Holographic calls
- Avatar robotics applications
- Nanonetworks

Toward 6G: A New Era of Convergence, First Edition. Amin Ebrahimzadeh and Martin Maier.
© 2021 The Institute of Electrical and Electronics Engineers, Inc.
Published 2021 by John Wiley & Sons, Inc.

- Flying networks
- Teleoperated driving (ToD)
- Electronic health (e-Health)
- Tactile Internet
- Internet of skills (IoS).

As a consequence, the network traffic will increase significantly with these new applications that will be enabled by technologies like virtual reality (VR) and augmented reality (AR). Even more exciting will be the widespread use and distribution of avatars for the reproduction and implementation of user actions. According to Yastrebova et al. (2018), avatar robotics applications can become one of the most important sources of traffic in future FG NET-2030 networks, involving new types of communications such as human-to-avatar (H2A), avatar-to-human (A2H), and avatar-to-avatar (A2A) communications. Importantly, taking into account the limited speed of propagation of light, the requirements for ultra-low latency should lead to the decentralization of future networks.

In academia, researchers from the University of Oulu's Centre for Wireless Communications launched an eight-year research program called *6G enabled smart society and ecosystem (6Genesis)* to conceptualize 6G. The first open 6Genesis seminar was held in August 2018. In Katz et al. (2018), an initial vision of what the sixth generation mobile communication system might be was presented by outlining the primary ideas of the 6Genesis Flagship Program (6GFP) created by the University of Oulu together with a Finish academic and industrial consortium. In this 6GFP program, 6G is investigated from a wide and realistic perspective, considering not only the communicational part of it but also looking into other highly relevant parts such as computer science, engineering, electronics, and material science. This integral approach is claimed to be instrumental in achieving truly novel solutions. Among others, the interrelated research areas of 6GFP aim at achieving distributed intelligent wireless computing by means of mobile edge, cloud, and fog computing. More specifically, intelligent distributed computing and data analytics is becoming an inseparable part of wireless networks, which call for self-organizing solutions to provide strong robustness in the event of device and link failures. Furthermore, VR/AR over wireless is considered one of the key application drivers for the future, whereby the information theory and practical performance requirements from the perspective of human psychology and physiology must be accounted for. As a consequence, perception-based coding should be considered to mitigate the shortcomings of existing compression–decompression algorithms in VR/AR. Future applications need distributed high-throughput local computing nodes and ubiquitous sensing to enable intelligent cyber-physical systems that are critical for future smart societies. Finally, techno-economic and business considerations need to address the question how network ownership and service provisioning models affect the design of radio access systems, including the potential analysis of high-risk technology enablers such as quantum theory and communications.

In September 2019, the world's first *6G white paper* was published as an outcome of the first 6G wireless summit, which was held in Levi, Finland, earlier in March 2019 with almost 300 participants from 29 countries, including major infrastructure manufacturers, operators, regulators as well as academia (Latva-aho and Leppänen, 2019). Each year, the white paper will be updated following the annual 6G wireless summit. While 5G was primarily developed to address the anticipated capacity growth demand from consumers and to enable the increasing importance of the IoT, 6G will require a substantially more holistic approach, embracing a much wider community. Many of the key performance indicators (KPIs) used for 5G are valid also for 6G. However, in the beyond 5G (B5G) and 6G, KPIs in most of the technology domains once again point to an increase by a factor of 10–100, though a 1000 times price reduction from the customer's view point may be also key to the success of 6G (Zhang et al., 2020). Note that price reduction is particularly important for providing connectivity to rural and underprivileged areas, where the cost of backhaul deployment is the major limitation. According to Yaacoub and Alouini (2020), providing rural connectivity represents a key 6G challenge and opportunity given that around half of the world population lives in rural or underprivileged areas. Among other important KPIs, 6G is expected to be the first wireless standard exceeding a peak throughput of 1 Tbit/s per user. Furthermore, 6G needs a network with embedded trust given that the digital and physical worlds will be deeply entangled by 2030. Toward this end, blockchain also known as distributed ledger technology (DLT) may play a major role in 6G networks due to its capability to establish and maintain trust in a distributed fashion without requiring any central authority.

Arguably more interestingly, the 6G white paper envisions that totally new services such as telepresence, as a surrogate for actual travel, will be made possible by combinations of graphical representations (e.g. avatars), wearable displays, mobile robots and drones, specialized processors, and next-generation wireless networks. Similarly, smartphones are likely to be replaced by pervasive extended reality (XR) experiences through lightweight glasses, whereby feedback will be provided to other senses via earphones and haptic interfaces.

1.2 Evolution of Mobile Networks and Internet

The general evolution of global mobile network standards was first to maximize coverage in the first and second generations and then to maximize capacity in the third and fourth generations. In addition to higher capacity, research on 5G mobile networks has focused on lower end-to-end latency, higher spectral efficiency and energy efficiency, and more connection nodes (Rowell and Han, 2015). More specifically, the first generation (1G) mobile network was designed for voice services with a data rate of up to 2.4 kbit/s. It used analog signal to transmit information, and there was no universal wireless standard. Conversely, 2G was based on digital modulation technologies and offered data rates of up to 384 kbit/s, supporting not only voice services but also data services such as short message service (SMS). The dominant 2G standard was the global system for mobile (GSM) communication. The third generation (3G) mobile network provided a data rate of at least 2 Mbit/s and enabled advanced services, including web browsing, TV streaming, and video services. For achieving global roaming, 3GPP was established to define technical specifications and mobile standards. 4G mobile networks were introduced in the late 2000s. 4G is an all Internet Protocol (IP) based network, which is capable of providing high-speed data rates of up to 1 Gbit/s in the downlink and 500 Mbit/s in the uplink in support of advanced applications like digital video broadcasting (DVB), high-definition TV content, and video chat. LTE-Advanced (LTE-A) has been the dominant 4G standard, which integrates techniques such as coordinated multipoint (CoMP) transmission and reception, multiple-input multiple-output (MIMO), and orthogonal frequency division multiplexing (OFDM). The main goal of 5G has been to use not only the microwave band but also the millimeter-wave (mmWave) band for the first time in order to significantly increase data rates up to 10 Gbit/s. Another feature of 5G is a more efficient use of the spectrum, as measured by increasing the number of bits per hertz. ITU's International Mobile Telecommunications 2020 (IMT 2020) standard proposed the following three major 5G usage scenarios: (i) enhanced mobile broadband (eMBB), (ii) ultra-reliable and low latency communications (URLLC), and (iii) massive machine type communications (mMTC). As 5G is entering the commercial deployment phase, research has started to focus on 6G mobile networks, which are anticipated to be deployed by 2030 (Huang et al., 2019).

Typically, next-generation systems do not emerge from the vacuum, but follow the industrial and technological trends from previous generations. Potential research directions of 6G consistent with these trends were provided by Bi (2019), including among others:

- *6G will continue to move to higher frequencies with wider system bandwidth*: Given that the spectrum at lower frequencies has almost been depleted, the current trend is to obtain wider bandwidth at higher frequencies in order to increase the data rate more than 10 times for each generation.
- *Massive MIMO will remain a key technology for 6G*: Massive MIMO has been the defining technology for 5G that has enabled the antenna number to increase from 2 to 64. Given that the performance gains have saturated in the areas of

channel coder and modulator, the hope of increasing spectral efficiency for 6G will remain in the multiple antenna area.

- *6G will take the cloud service to the next level*: With the ever higher data rates, short delays, and low transmission costs, many of the computational and storage functions have been moved from the smartphone to the cloud. As a result, most of the computational power of the smartphone can focus on presentation rendering, making VR, AR, or XR more impressive and affordable. Many artificial intelligence (AI) services that are intrinsically cloud based may prevail more easily and broadly. In addition to smartphones, less expensive functional terminals may once again flourish, providing growth opportunities in more application areas.
- *Grant-free transmissions could be more prominent in 6G*: In past cellular network generations, transmissions were primarily based on grant-oriented design with strong centralized system control. More advanced grant-free protocols and approaches will be needed for 6G. It is possible that the non-orthogonal multiple access (NOMA) technology may have another opportunity to prevail due to its short delay performance even though it failed to take off during the 5G time period.
- *mMTC is more likely to take shape in the older generation before it can succeed in the next generation*: mMTC has been one of the major directions for the next-generation system design since the market growth of communications between people has saturated. High expectations have been put on 5G mMTC to deliver significant growth for the cellular industry. Until now, however, this expectation has been mismatched with the reality on the ground. Therefore, the current trend appears to indicate that mMTC would be more likely to prevail by utilizing older technology that operates in the lower band at lower cost.
- *6G will transform a transmission network into a computing network*: One of the possible trademarks of 6G could be the harmonious operations of transmission, computing, AI, machine learning, and big data analytics such that 6G is expected to detect the users' transmission intent autonomously and automatically provide personalized services based on a user's intent and desire.

In his latest book "The Inevitable," Kevin Kelly described the 12 technological forces that will shape our future (Kelly, 2016). According to Kelly, nothing has happened yet in terms of the Internet. The Internet linked humans together into one very large thing. From this embryonic net will be born a collaborative interface, a sensing, cognitive apparatus with power that exceeds any previous invention. The hard version of it is a future brought about by the triumph of a superintelligence. According to Kelly, however, a soft singularity is more likely where AI and robots converge – humans plus machines – and together we move to a complex interdependence. This phase has already begun. We are connecting all humans and all machines into a global matrix, which some call the global mind or world brain. It is a new regime wherein our creations will make us better humans. This new platform will include the collective intelligence of all humans combined with the collective behavior of all machines, plus the intelligence of nature, plus whatever behavior emerges from this whole. Kelly estimates that by the year 2025 every person will have access to this platform via some almost-free device.

The importance of convergence of emerging key technologies, e.g. AI, robots, and XR, lies also at the heart of the 6G era with standards and enabled devices anticipated to roll out around 2030. 6G research is just now starting, even though 5G networks have not been widely deployed yet. A few countries, most notably Finland as well as China and South Korea, have taken the lead by launching 6G programs to avoid getting left behind.

1.3   6G Network Architectures and Key Enabling Technologies

1.3.1   Four-Tier Networks: Space-Air-Ground-Underwater

6G network architectures are anticipated to extend the 5G three-tier space-air-ground networks by integrating underwater networks, thus giving rise to four-tier space-air-ground-underwater networks with near-instant and unlimited superconnectivity in the sky, at sea, and on land. According to Zhang et al. (2019b), these large-dimensional integrated nonterrestrial and terrestrial networks will consist of the following four network tiers:

- *Space-network tier*: This network tier will support orbit or space Internet services in such applications such as space travel and provide wireless coverage via satellites. For long-distance intersatellite transmission in free space, laser communications represents a promising solution. The use of mmWave frequencies to establish high-capacity (inter)satellite communications may be another feasible solution to complement terrestrial 6G networks with computing stations placed on satellite platforms (Giordani and Zorzi, 2020). The integration of terrestrial and non-terrestrial networks poses a number of challenges and new open problems such as (i) large propagation delays, (ii) Duppler effect due to fast moving satellites, and (iii) severe path loss of mmWave transmission.
- *Air-network tier*: This network tier works in the low-frequency, microwave, and mmWave bands to provide more flexible and reliable connectivity for urgent events or in remote areas by densely employing flying base stations, e.g. unmanned aerial vehicles (UAVs).
- *Terrestrial-network tier*: Similar to 5G, this network tier will still be the main solution for providing wireless coverage for most human activities. It will

support low-frequency, microwave, mmWave, and THz bands in ultradense heterogeneous networks, which require the deployment of ultra-high-capacity backhaul infrastructures. Optical fiber will still be important for 6G, though THz wireless backhaul will be an attractive alternative.

- *Underwater-network tier*: Finally, this network tier will provide coverage and Internet services for broad-sea and deep-sea activities for military or commercial applications. Given that water exhibits different propagation characteristics, acoustic and laser communications can be used to achieve high-speed data transmission for bidirectional underwater communications. According to Huang et al. (2019), however, there is a lot of controversy about whether undersea networks are able to become a part of future 6G networks. Unpredictable and complex underwater environments lead to intricate network deployments, severe signal attenuation, and physical damage to equipment, leaving plenty of issues to be resolved.

1.3.2 Key Enabling Technologies

1.3.2.1 Millimeter-Wave and Terahertz Communications

Higher frequencies from 100 GHz to 3 THz are promising bands for the next generation of wireless communication systems, offering the potential for revolutionary applications. Technically, the formal definition of the THz region is 300 GHz through 3 THz, though sometimes the terms sub-THz or sub-mmWave are used to define the 100–300 GHz spectrum. The short wavelengths at mmWave and THz will allow massive spatial multiplexing in hub and backhaul communications. The THz band from 100 GHz through 3 THz can enable secure communications due to the fact that small wavelengths allow for extremely high-gain antennas with extremely small physical dimensions. The ultra-high data rates facilitated by mmWave and THz wireless local area and cellular networks will enable super-fast download speeds for computer communication, autonomous vehicles, robotic control, the so-called information shower, high-definition holographic gaming, and high-speed wireless data distribution in data centers. In addition to the extremely high data rates, there are promising applications for future mmWave and THz systems that are likely to evolve in 6G networks and beyond. These applications can be categorized into the main areas of wireless cognition, sensing, imaging, wireless communications, and position location/THz navigation (Rappaport et al., 2019).

A comprehensive literature review on the technical challenges in THz communications for B5G wireless networks was presented by Chen et al. (2019). In this survey, several key technologies for the realization of THz wireless communication systems were discussed in technically greater detail. Heterodyne reception is the most widespread receiving system in the THz band, whereby its core circuits usually include the circuits for frequency conversion, signal generation, and amplification. In the THz band, however, solid state amplifiers are lacking because the technology of compound semiconductor transistors is immature. Further, due to the lack of THz amplifiers, mixers become the first stage of receivers and affect their system performance. In the THz band, subharmonic mixers are usually used because they can mitigate the difficulty of local oscillators. The combination of metamaterials and semiconductor technologies has led to significant breakthroughs in dynamic THz functional devices, including THz amplitude and phase modulation. For the sake of channel characterization and propagation measurements in future THz wireless communication systems, it is vital to establish efficient channel models that maximize THz bandwidth allocation and spectral efficiency. Channel estimation in THz communication systems is challenging due to hybrid beamforming structures and the large number of antennas. Large-scale phased array antennas are suitable for THz communication systems to compensate for the high path loss and molecular absorption loss.

1.3.2.2 Reconfigurable Intelligent Surfaces

A brand-new wireless communication technology referred to as reconfigurable intelligent surfaces (RISs) – also known as large intelligent surfaces, smart reflect-arrays, intelligent reflecting surfaces, passive intelligent mirrors, artificial radio space, or programmable metasurface – has emerged recently (Basar et al., 2019; Tang et al., 2020b). RISs are often referred to as software-defined surfaces (SDSs) in analogy with the concept of software-defined radio (SDR). Accordingly, an RIS may be viewed as an SDS whose surface of electromagnetic material is controlled with integrated electronics and its response of the radio waves is programmed in software.

According to Basar et al. (2019), the distinctive characteristic of RISs lies in making the environment controllable by the telecommunication operators and thus giving them the possibility of shaping and fully controlling the electromagnetic response of the environmental objects that are distributed throughout the network. As a result, network operators are able to control the scattering, reflection, and refraction characteristics of the radio wave and thereby effectively control the wavefront (e.g. phase, amplitude, frequency, and even polarization) of wireless signals without the need of complex decoding, encoding, and radio frequency processing operations. In contrast to conventional wireless networks, where the environment is out of control of the telecommunication operators, RISs render the wireless environment a smart reconfigurable space that plays an active role in transferring and processing information. Consequently, RISs have given rise to the emerging concept of smart radio environments. In smart radio environments, the wireless environment is turned into a software-reconfigurable entity, whose operation is optimized to enable uninterrupted connectivity and high QoS guarantees. This is in stark contrast to conventional wireless networks, where the radio environment has usually an uncontrollable negative effect on the communication efficiency and QoS due to signal attenuation, multipath propagation, fading, and reflections from objects. RISs have the following distinguishable features (Basar et al., 2019):

- They are nearly passive and, ideally, do not need any dedicated energy source.
- They form a contiguous surface and, ideally, any point can shape the wave impinging upon it.
- They are not affected by receiver noise since, ideally, they do not need analog-to-digital converter (ADCs)/digital-to-analog converter (DACs) and power amplifiers.
- They have full-band response since, ideally, they can work at any operating frequency.
- They can be easily deployed, e.g. on facades of building, ceilings of indoor spaces, or human clothing.

1.3.2.3 From Network Softwarization to Network Intelligentization

In contrast to previous generations, 6G will be transformative and will revolutionize the wireless evolution from "connected things" to "connected intelligence." According to Letaief et al. (2019), 6G will take network softwarization to a new level, namely toward network intelligentization. Software-defined networking (SDN) and network function virtualization (NFV) have moved modern communications networks toward software-based virtual networks. They also enable network slicing, which can provide a powerful virtualization capability to allow multiple virtual networks to be created atop a shared physical infrastructure. However, as the network is becoming more complex and more heterogeneous, softwarization is not going to be sufficient for 6G. Existing technologies such as SDN, NFV, and network slicing will need to be further improved by enabling fast learning and adaptation via AI-based methods. As a result, network slicing will become much more versatile and intelligent in order to support diverse capabilities and more advanced IoT functionalities, including sensing, data collection, analytics, and storage.

6G is expected to undergo an unprecedented transformation that will make it substantially different from the previous generations of wireless cellular systems. In particular, 6G will go beyond mobile Internet and will be required to support ubiquitous AI services from the core to the end devices of the network. Toward this end, Letaief et al. (2019) argue that 6G will require the support of the following three new service types beyond the aforementioned 5G eMBB, URLLC, and mMTC services:

- *Computation oriented communications (CoC)*: New smart devices call for distributed computation to enable key functionalities such as federated learning. Instead of targeting conventional QoS provisioning, computation oriented communication (CoC) will flexibly choose an operating point in the rate-latency-reliability space depending on the availability of various communications resources to achieve a certain computational accuracy.
- *Contextually agile eMBB communications (CAeC)*: The provision of 6G eMBB services is expected to be more agile and adaptive to the network context, including the communication network context such as link congestion and network topology, the physical environment context such as surrounding location and mobility, and the social network context such as social neighborhood and sentiments.
- *Event defined URLLC (EDURLLC)*: In contrast to the 5G URLLC application scenario with redundant resources in place to offset many uncertainties, 6G event defined uRLLC (EDURLLC) will need to support URLLC in extreme or emergency events with spatially and temporally changing device densities, traffic patterns, and spectrum and infrastructure availability.

6G will provide an information and communication technology (ICT) infrastructure that enables end users to perceive themselves as surrounded by a huge artificial brain offering virtually zero-latency services, unlimited storage, and immense cognition capabilities. 6G will play a significant role in responding to fundamental human and social needs and in helping realize Nikola Tesla's prophecy that "when wireless is perfectly applied, the whole Earth will be converted into a huge brain", according to Strinati et al. (2019). Toward this end, however, network intelligentization still has a long way to go by advancing machine learning technologies for 6G by taking more KPIs different from the traditional metrics into account, including situational awareness, learning ability, storage cost, and computation capacity (Kato et al., 2020). This also applies to the future intelligentization of 6G vehicular networks, where employing machine learning in vehicular communications becomes a hot topic that is widely studied in both academia and industry (Tang et al., 2020a).

When it comes to defining the unique challenges and opportunities of 6G, it is important to note that there is a strong notion that the nature of mobile terminals will change, with cars and mobile robots playing a more important role. Furthermore, we might witness the union of network convergence, meaning that we may see stronger dependencies between networking infrastructures and applications (David et al., 2019).

1.4 Toward 6G: A New Era of Convergence

According to the authors of Saad et al. (2020), the current deployment of 5G cellular systems is exposing the inherent limitations of this system, compared to its original premise as an enabler for Internet of everything (IoE) applications. IoE services will require an end-to-end design of communication, control, and computation functionalities, which to date has been largely overlooked. These 5G drawbacks are currently spurring worldwide activities focused on defining the next-generation 6G wireless system that can truly integrate far-reaching applications ranging from autonomous systems to XR and haptics. Importantly, the authors opine that 6G will not be a mere exploration of more spectrum at high-frequency bands, but it will rather be a *convergence of upcoming technological trends*. Toward this end, the authors presented a holistic, comprehensive research agenda that leverages those technologies and serves as a basis for stimulating more out-of-the-box research around 6G. While traditional applications will remain central to 6G, the key determinants of the system performance will be the following four new applications domains: (i) multisensory XR applications, (ii) connected robotics and autonomous systems, (iii) wireless brain-computer interaction, a subclass of human–machine interaction (HMI), and (iv) blockchain and distributed ledger technologies.

In addition to many of the 6G driving trends and enabling technologies discussed in previous sections, Saad et al. (2020) emphasized the importance of haptic and empathic communications and the emergence of new human-centric service classes as well as the *end of the smartphone era*. They argue that smartphones were central to 4G and 5G. However, in recent years there has been an increase in wearable devices whose functionalities are gradually replacing those of smartphones. This trend is further fueled by applications such as XR and HMI, e.g. brain-computer interaction. The devices associated with those applications range from smart wearables to integrated headsets or even smart body implants that can take direct sensory inputs from human senses, bringing an end to smartphones and potentially driving a majority of 6G use cases. They also expect that a handful of technologies will mature along the same time of 6G, e.g. quantum computing and communications, and hence potentially play a role toward the end of the 6G standardization and research process.

An interesting example of out-of-the-box 6G research was presented just recently in Viswanathan and Mogensen (2020). The authors claim that new themes are likely to emerge. Specifically, the future of connectivity is in the creation of *digital twin worlds* that are a true representation of the physical and biological worlds at every spatial and time instant, unifying our experience across these physical, biological, and digital worlds. Digital twins of various objects created in edge clouds will form the essential foundation of the future digital world. Digital twin worlds of both physical and biological entities will be an essential platform for the new digital services of the future. Digitalization will also pave the way for the creation of new virtual worlds with digital representations of imaginary objects that can be blended with the digital twin world to various degrees to create a mixed-reality, super-physical world. Smart watches and heart rate monitors will be mapped accurately every instant and integrated into the digital and virtual worlds, enabling new *super-human capabilities*. AR user interfaces will enable efficient and intuitive human control of all these worlds, whether physical, virtual, or biological, thus creating a unified experience for humans and the human transformation resulting from it. Dynamic digital twins in the digital world with increasingly accurate, synchronous updates of the physical world will be an essential platform for augmenting human intelligence.

The authors of Viswanathan and Mogensen (2020) outlined a vision of the future life and digital society on the other side of the 2030s. While the smartphone and the tablet will still be around, we are likely to see new man–machine interfaces that will make it substantially more convenient for us to consume and control information. The authors expect that wearable devices, such as earbuds and devices embedded in our clothing, will become common. We will have multiple wearables that we carry with us and they will work seamlessly with each other, providing natural, intuitive interfaces. Touch-screen typing will likely become outdated. Gesturing and talking to whatever devices we use to get things done will become the norm. The devices we use will be fully context-aware and the network will become increasingly sophisticated at predicting our needs. This context awareness combined with new man–machine interfaces will make our interaction with the physical and digital world much more intuitive and efficient. The computing needed for these devices will likely not all reside in the devices themselves because of form factor and battery power considerations. Rather, they may have to rely on locally available computing resources to complete tasks beyond the edge cloud. As consumers, we can expect that the *self-driving concept cars* of today will be available to the masses by the 2030s. They will be self-driving most of the time and thus will substantially increase the time available for us to consume data from the Internet in the form of more entertainment, rich communications, or education. Further, numerous *domestic service robots* will complement the vacuum cleaners and lawn mowers we know today. These may take the form of a swarm of smaller robots that work together to accomplish tasks.

Combining the multi-modal sensing capabilities with the cognitive technologies enabled by the 6G platform will allow for analyzing behavioral patterns and people's preferences and even emotions, hence creating a sixth sense that anticipated user needs. The resultant *network with the sixth sense* will allow for interactions with the physical world in a much more intuitive way.

1.5   Scope and Outline of Book

1.5.1   Scope

Building on the 6G vision outlined above, this book will describe the latest developments and recent progress on the key technologies enabling next-generation 6G mobile networks, paying particular attention to their seamless convergence. To help make and keep things concrete, the book will focus on the emerging Tactile Internet as one of the most interesting 5G/6G URLLC applications. Beside conventional audiovisual and data traffic, the Tactile Internet envisions the real-time transmission of haptic information (i.e. touch and actuation) for the remote control of physical and/or virtual objects through the Internet. The Tactile Internet opens up a plethora of exciting research directions toward adding a new dimension to the human-to-machine interaction via the Internet by exploiting context- as well as self-awareness. The underlying end-to-end design approach of the Tactile Internet is fully reflected in the key principles of the Tactile Internet. Among others, the key principles envision to support local area as well as wide area connectivity through wireless or hybrid wireless/wired networking. Furthermore, it leverages computing resources from cloud variants at the edge of the network. Some of the key use cases of the Tactile Internet include teleoperation, haptic communications, immersive VR, and automotive control. We will leverage our expertise and extend our recent work on immersive Tactile Internet experiences in unified fiber-wireless mobile networks based on AI enhanced multi-access edge computing (MEC), including cooperative computation offloading.

In addition, we will include our work on decentralizing the Tactile Internet in general and edge computing in particular via Ethereum blockchain technologies, most notably the so-called decentralized autonomous organization (DAO). Unlike AI-based agents that are completely autonomous, a DAO still requires heavy involvement from humans specifically interacting according to a protocol defined by the DAO in order to operate. We will elaborate on how this particular feature of DAOs (i.e. automation at the center and humans at the edges) can be exploited in the emerging concept of human-agent-robot teamwork.

Finally, we report on the state-of-the-art and our ongoing work on XR in the post-smartphone era. Specifically, we will elaborate on the implications of the transition from the current gadgets-based Internet to a future Internet that is evolving from bearables (e.g. smartphone), moves toward wearables (e.g. Google and Levi's smart jacket or Amazon's recently launched voice-controlled Echo Loop ring, glasses, and earbuds), and then finally progresses to nearables (e.g. intelligent mobile robots). Nearables denote nearby surroundings or environments with embedded computing/storage technologies and service provisioning mechanisms that are intelligent enough to learn and react according to user context and history in order to provide user-intended services. While 5G was supposed to be about the IoE, to be transformative 6G might be just about the opposite of Everything, i.e. Nothing or, more technically, No Things. Toward this end, we will elaborate on the *Internet of No Things* as an extension of immersive VR from virtual to real environments, where human-intended Internet services – either digital or physical – appear when needed and disappear when not needed. Building on Nissan's so-called invisible-to-visible (I2V) technology concept for self-driving cars, we will explore how the full potential of multisensory XR experiences may be unleashed in so-called Multiverse cross-reality environments and present our *extrasensory perception network (ESPN)* for the nonlocal extension of human "sixth-sense" experiences in space and time.

1.5.2 Outline

The remainder of the book comprises the following six chapters:

In Chapter 2, we elaborate on the Tactile Internet and its inherent human-in-the-loop (HITL) nature of human-to-machine interaction, paying close attention to the dichotomy between automation and augmentation (i.e. extension of capabilities) of the human. The Tactile Internet allows for a human-centric design approach toward creating novel immersive experiences and extending the capabilities of the human through the Internet by means of haptic communications and teleoperation. In this chapter, we pay attention to bilateral teleoperation as an example of HITL-centric applications and present an in-depth study of haptic traffic characterization and modeling. Specifically, we develop models of packet interarrival times and three-dimensional sample autocorrelation based on haptic traces obtained from real-world teleoperation experiments. Furthermore, we explore how wireless edge intelligence can be leveraged to help realize immersive teleoperation experiences in mobile networks that are unified with fiber backhaul and wireless mesh front-end networks based on low-cost data-centric optical fiber Ethernet, i.e. Ethernet passive optical network (EPON), and wireless Ethernet, i.e. wireless local area network (WLAN), technologies.

In Chapter 3, with the rise of increasingly smarter machines, we explore coworking with mobile robots – owned by mobile users (i.e. ownership spreading) or the mobile network operator – in greater detail by shedding light on the coordination of the human–robot symbiosis. A promising approach toward achieving advanced human–machine coordination by means of a superior process for fluidly orchestrating human and machine coactivity, which may vary over time or be unpredictable in different situations, can be found in the still young field of human-agent-robot-teamwork (HART) research. Toward this end, we investigate how context-awareness may be used to develop a HART-centric multi-robot task coordination algorithm that minimizes the completion time of physical and digital tasks as well as operational expenditures (OPEX) by spreading ownership of robots across mobile users. In addition, we explore how self-awareness can be exploited to improve the performance of multiple robots by identifying their respective capabilities as well as the objective requirements by means of optimal motion planning to minimize their energy consumption and traverse time to given physical and/or digital tasks. The proposed context- and self-aware HART-centric allocation scheme for both physical and digital tasks may be used to coordinate the automation and augmentation of mutually beneficial human–machine coactivities across the Tactile Internet based on unified communication network infrastructures.

In Chapter 4, we delve into the so-called missing middle that refers to the new ways that have to bridge the gap between human-only and machine-only activities for creating cutting-edge jobs and innovative businesses. This gives way to the so-called third wave of business transformation, which will be centered around human + machine activities. Toward this end, we formulate and solve the problem of joint prioritized scheduling and assignment of delay-constrained teleoperation tasks to available skilled human operators across unified communication network infrastructures with multiple objectives to minimize the average weighted task completion time, maximum tardiness, and average OPEX per task. We develop an analytical framework to estimate the end-to-end delay of both local and nonlocal teleoperation across the enhanced mobile networks under consideration and investigate the coexistence of conventional human-to-human (H2H) and haptic human-to-machine (H2M) traffic.

In Chapter 5, we explore the beneficial impact of cooperative computation offloading on the quality of experience (QoE) of mobile users with regard to average response time between mobile users, MEC servers, and remote cloud. Specifically, we investigate techniques that enable mobile users in self-organizing cellular networks to adaptively adjust their computational speed in order to reduce energy consumption or shorten task execution time under different scenarios. In our design approach, we take into account limitations stemming from both communications and computation by accurately modeling the fronthaul/backhaul as well as edge/cloud servers, while paying particular attention to the offloading decision making between mobile users and edge servers as well as edge servers and remote cloud. To allow mobile users to flexibly rely on their local computing resources by means of dynamic reconfiguration, the proposed self-organization framework lets mobile devices tune their offloading probability and computational capabilities adaptively, thus giving rise to a Pareto frontier characterization of the trade-off between average task execution time and energy consumption.

In Chapter 6, we explore the salient features that set Ethereum aside from other blockchains in more depth, including their symbiosis with other emerging key technologies such as AI and robots apart from blockchain-enabled edge computing. A question of particular interest hereby is how decentralized blockchain mechanisms – most notably Ethereum's concept of the DAO – may be leveraged to let emerge new hybrid forms of collaboration among individuals, which havenot been entertained in the traditional market-oriented economy dominated by firms rather than individuals. After elaborating on the commonalities of and specific differences between Ethereum and Bitcoin blockchains, we explain DAO in more detail and discuss the potential role of Ethereum and in particular the DAO in helping decentralize the Tactile Internet as a promising example of future techno-social systems via automation at the center and crowdsourcing of human assistance at the edges. Further, we explore the possibilities to extend the smart contract framework of the emerging blockchain Internet of things (BIoT) for enabling the nudging of human users in a broader Tactile Internet context by searching for synergies between the aforementioned HART and the complementary strengths of the DAO, AI, and robots.

Finally, in Chapter 7, we take an outlook on how future profound 6G technologies will weave themselves into the fabric of everyday life until they are indistinguishable from it. In our discussion, we show that future fully interconnected VR systems and the Tactile Internet seem to evolve toward common design goals. Most notably, the boundary between virtual (i.e. online) and physical (i.e. offline) worlds is to become increasingly imperceptible, while both digital and physical capabilities of humans are to be extended via edge computing variants with embedded AI capabilities. More specifically, we elaborate on the far-reaching vision of future 6G networks ushering in an anticipated 6G post-smartphone era, where smartphones will be increasingly replaced with wearables (e.g. smart jackets or voice-controlled glasses/earbuds/rings) and nearables (e.g. intelligent mobile robots). After explaining the reality–virtuality continuum in more detail, we introduce the so-called Multiverse to unleash the full potential of advanced XR technologies for the extension of human experiences, ranging from conventional VR to more sophisticated cross-reality environments known as third spaces. Further, we explore the potential of the recently emerging I2V technology concept, which we use together with other key enabling technologies (AI enhanced MEC, intelligent mobile robots, blockchain) to tie both online and offline worlds closer together in order to make the enduser "see the invisible" through the awareness of nonlocal events in space and time by mimicking the quantum realm via emerging multisensory XR and extrasensory "sixth-sense" human experiences.

2

Immersive Tactile Internet Experiences via Edge Intelligence

2.1 Introduction

Beside conventional audiovisual and data traffic, the emerging *Tactile Internet* envisions the real-time transmission of haptic information (i.e. touch and actuation) for the remote control of physical and/or virtual objects through the Internet (Simsek et al., 2016). The Tactile Internet holds promise to provide a paradigm shift in how skills and labor are digitally delivered globally, thereby converting today's content-delivery networks into skillset/labor-delivery networks (Aijaz et al., 2017). The Tactile Internet is expected to have a profound socioeconomic impact on a broad array of applications in our everyday life, ranging from industry automation and transport systems to healthcare, telesurgery, and education. Toward this end, at the core of the design of the Tactile Internet is realizing the so-called <10 ms-challenge (i.e. achieving a round-trip latency of <10 ms) with carrier-grade reliability.

The term "Tactile Internet" was first coined by G. P. Fettweis in 2014. In his seminal paper, Fettweis (2014) defined the Tactile Internet as a breakthrough enabling unprecedented mobile applications for tactile steering and control of real and virtual objects by requiring a round-trip latency of 1–10 ms. Later in 2014, ITU-T published a Technology Watch Report on the Tactile Internet, which emphasized that scaling up research in the area of wired and wireless access networks will be essential, ushering in new ideas and concepts to boost access networks' redundancy and diversity to meet the stringent latency as well as carrier-grade reliability requirements of Tactile Internet applications (ITU-T Technology Watch Report, 2014).

Toward 6G: A New Era of Convergence, First Edition. Amin Ebrahimzadeh and Martin Maier.
© 2021 The Institute of Electrical and Electronics Engineers, Inc.
Published 2021 by John Wiley & Sons, Inc.

To give it a more 5G-centric flavor, the Tactile Internet has been more recently also referred to as the 5G-enabled Tactile Internet (Aijaz et al., 2017 and Simsek et al., 2016). Unlike the previous four cellular generations, 5G networks will lead to an increasing integration of cellular and WiFi technologies and standards (Andrews et al., 2014). Furthermore, the importance of the so-called *backhaul bottleneck* needs to be recognized as well, calling for an end-to-end design approach leveraging both wireless front-end and wired backhaul technologies. Or, as eloquently put by J. G. Andrews, the lead author of Andrews et al. (2014), "placing base stations all over the place is great for providing the mobile stations high-speed access, but does this not just pass the buck to the base stations (BSs), which must now somehow get this data to and from the wired core network?" (Andrews 2013).

This mandatory end-to-end design approach is fully reflected in the key principles of the reference architecture within the emerging IEEE P1918.1 standards working group (formed in March 2016), which aims to define a framework for the Tactile Internet (Aijaz et al., 2018). Among others, the key principles envision to (i) develop a generic Tactile Internet reference architecture, (ii) support local area as well as wide area connectivity through wireless (e.g. cellular, WiFi) or hybrid wireless/wired networking, and (iii) leverage computing resources from cloud variants at the edge of the network. The working group defines the Tactile Internet as follows: "A network, or a network of networks, for remotely accessing, perceiving, manipulating or controlling real and virtual objects or processes in perceived real-time." Some of the key use cases considered in IEEE P1918.1 include teleoperation, haptic communications, immersive virtual reality (VR), and automotive control.

Clearly, the Tactile Internet opens up a plethora of exciting research directions toward adding a new dimension to the human-to-machine (H2M) interaction via the Internet. According to the aforementioned ITU-T Technology Watch Report, the Tactile Internet is supposed to be the next leap in the evolution of today's Internet of thing (IoT), though there is a significant overlap among 5G, IoT, and the Tactile Internet, as illustrated in Figure 2.1. Despite their differences, all three share an intersecting set of design goals:

- Very low latency on the order of 1 ms
- Ultrahigh reliability with an almost guaranteed availability of 99.999%
- Human-to-human (H2H)/machine-to-machine (M2M) coexistence
- Integration of data-centric technologies with a particular focus on WiFi
- Security.

For illustration, Figure 2.2 depicts a typical teleoperation system based on bidirectional haptic communications between a human operator (HO) and a teleoperator robot (TOR). Note that the number of independent coordinates required to

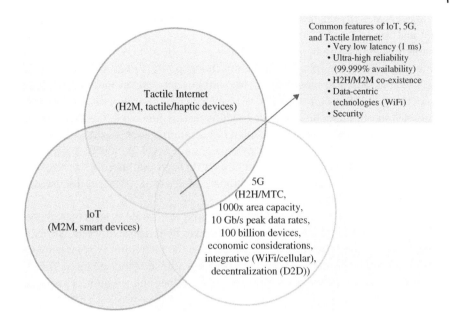

Figure 2.1 The three lenses of 5G, Internet of things (IoT), and the Tactile Internet: Commonalities and differences. Source: Maier et al. (2016). © 2016 IEEE.

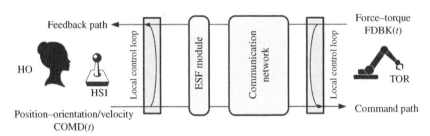

Figure 2.2 Teleoperation system based on bidirectional haptic communications between a human operator (HO) and a teleoperator robot (TOR). Source: Maier and Ebrahimzadeh (2019). © 2019 IEEE.

completely specify and control/steer the position, orientation, and velocity of the TOR is defined by its degrees-of-freedom (DoF)[1]. Further, a local human system interface (HSI) device is used to display haptic interaction with the remote TOR to the HO. The local control loops on both ends of the teleoperation system ensure the tracking performance and stability of the HSI and TOR.

---

1 Currently available teleoperation systems range from 1-DoF to >20-DoF TORs. For instance, a 6-DoF TOR allows for both translational motion (in 3D space) via force and rotational motion (pitch, yaw, and roll) via torque.

Maier et al. (2016) elaborated on the subtle differences between the Tactile Internet and the IoT and 5G, which may be best expressed in terms of underlying communications paradigms and enabling end-devices. Importantly, the Tactile Internet involves the inherent human-in-the-loop (HITL) nature of H2M interaction, as opposed to the emerging IoT without any human involvement in its underlying M2M communications. While M2M communications is useful for the automation of industrial and other machine-centric processes, the Tactile Internet will be centered around H2M/robot (R) communication and thus allows for a human-centric design approach toward creating novel immersive experiences and extending the capabilities of the human through the Internet, i.e. augmentation rather than automation of the human (Maier et al., 2018), as discussed in more detail in Section 2.2.

Deep fiber access solutions have been deployed worldwide to push optical fiber closer to individual homes and businesses and to help realize different flavors of fiber-to-the-x (FTTx) networks, where x denotes the discontinuity point between optical fiber and some other wired or wireless transmission medium. Today's broadband access networks leverage both optical fiber and wireless technologies with seamless convergence, giving rise to bimodal fiber-wireless (FiWi) access networks (Maier and Ghazisaidi, 2018). FiWi access networks combine the reliability, robustness, and high capacity of optical fiber networks and the flexibility, ubiquity, and cost savings of wireless networks. Fully converged networks, where different fixed and mobile access technologies can be flexibly selected while sharing core network functionalities, will be instrumental in realizing *5G low-latency applications*. This is particularly advantageous for those use cases that do not necessarily require mobility all of the time and thus can be carried out in fixed broadband network environments (Lema et al., 2017).

Historically, the huge bandwidth potential of optical fiber, which is far in excess of any other utilized transmission medium, has lured most research efforts into focusing on the primary goal of continuously increasing the capacity of optical networks rather than on, for example, lowering their end-to-end latency. This comes as no surprise, given that a single strand of fiber offers a total bandwidth of 25 000 GHz, which can be easily tapped into using wavelength division multiplexing (WDM). To put this potential into perspective, it is worthwhile to note that it is about 1000 times the entire usable radio frequency (RF) spectrum on the planet Earth (Green, 2001). As an illustrative example, Figure 2.3 shows the next-generation passive optical network (NG-PON) roadmap as envisioned back in 2009, where the primary design goal for (r)evolutionary NG-PON1&2 broadband access networks was the provisioning of ever increasing capacity over time (Kani et al., 2009). However, this perspective slowly started to change in 2013,[2] when

---

2 OFC/NFOEC workshop on "Post NG-PON2: Is it More About Capacity or Something Else," 2013.

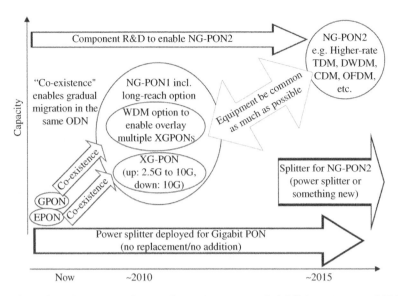

Figure 2.3 Next-generation passive optical network (NG-PON) roadmap as of 2009 illustrating the primary goal of continued capacity upgrades in the past. Source: Kani et al. (2009). © 2009 IEEE.

questions surfaced whether to focus access research efforts on more than just continued capacity upgrades (Maier, 2014). According to Biermann et al. (2013), one of the major factors limiting the performance of edge mobile networks is latency.

Passive optical network (PON) technologies are anticipated to accelerate 5G deployments (Pfeiffer, 2018). One solution to reduce latency in PONs is to modify the architectural structure of the remote node by adding loop-back fibers to the passive splitter. In doing so, a local *Fx fronthaul* (Fx-FH) can be realized for direct inter-optical network unit (ONU) communications, where ONUs may interface with their collocated macro and small-cell BSs, thus forming local clusters for coordinated multipoint (CoMP) transmission in long-term evolution-advanced (LTE-A) networks. Note that Fx is used to denote various lower-layer split points along the 5G radio processing chain, as specified by the ITU-T Supplement G.Sup.5GP (2018). Pfeiffer (2018) also emphasized the importance of end-to-end coordination of both PON and wireless network resources via a common orchestrator that runs one or more *cooperative dynamic bandwidth allocation (co-DBA)* algorithms in support of emerging 5G low-latency applications. A preliminary study of a distributed medium access control (MAC) protocol and simple dynamic bandwidth allocation (DBA) algorithm run by ONUs to support low-latency communication among BSs across multiple PONs, whose neighboring remote nodes were interconnected via additional fiber links, was presented by Li and

Chen (2017). This study exploited PON-based technologies for realizing mobile *backhaul* infrastructures. Similarly, in Ranaweera et al. (2013a,b), AT&T reported on their strategy to leverage existing PON-based fiber-to-the-node (FTTN) residential access, right of way, and already installed powering facilities to provide inexpensive small-cell backhaul. Conversely, mobile *fronthaul* networks, which interconnect centralized baseband units (BBUs) with remote radio heads (RRHs) located at cell sites, have been implemented by using digital fiber-optic interfaces such as the common public radio interface (CPRI) (Kim, 2018). For instance, China Mobile's cloud radio access network (C-RAN) is CPRI based and hence makes use of digital radio-over-fiber (RoF) techniques.

Clearly, one way to realize the common orchestrator is by centralizing all end-to-end coordination functions in the cloud, giving rise to the widely studied C-RAN; see, e.g. Velasco et al. (2017), Zhou et al. (2018), and Pérez et al. (2018). C-RAN is able to achieve significant cost savings by sharing centralized network resource management units among mobile users (MUs). We revisit C-RAN in Section 2.4 and elaborate on their pros and cons in light of the emerging concept of *edge computing* (Rimal et al., 2017c). Edge computing is a new paradigm in which computing and storage resources – variously referred to as cloudlets, micro datacenters, or fog nodes – are placed at the Internet's edge in proximity to wireless end devices in order to achieve low end-to-end latency, low jitter, and scalability (Satyanarayanan 2017).

In this chapter, we pay attention to bilateral teleoperation as an example of HITL applications and present an in-depth study of haptic traffic characterization and modeling in terms of packet arrival and sample autocorrelation. We develop new models of describing packet interarrival times as well as three-dimensional sample autocorrelation. We then explore how edge intelligence may be leveraged to help realize an immersive, reliable teleoperation experience over FiWi-based networking infrastructures. More specifically, we focus on the communication network in Figure 2.2 and its role in realizing the Tactile Internet vision, thereby paying particular attention to the unique characteristics of haptic traffic. According to Steinbach et al. (2012), even minor communication-induced time delays and packet losses may destabilize the haptic communications system. Emphasizing on its HITL-centric aspect, the Tactile Internet allows for a human-centric design approach by exploiting the properties of human haptic perception via advanced perceptual coding techniques in order to substantially reduce the haptic packet rate, as explained in technically greater detail shortly. The contributions of this chapter are threefold:

(i) First, we model Tactile Internet traffic by means of extensive haptic traces, taking TORs with different DoF and perceptual coding into account. As shown in Figure 2.2, in a typical teleoperation system the position-orientation/velocity samples are transmitted from the HO through the HSI in the command path, whereas the force–torque samples are sent back to the HO in the feedback path. In teleoperation, haptic feedback plays a crucial role in providing the HO with transparency, immersion, and togetherness with the remote environment (Steinbach et al., 2012). Note, however, that despite growing interest in the Tactile Internet, there is still limited understanding of the characteristics of real haptic traffic, especially at the packet level. For simplicity and analytical tractability, Tactile Internet traffic has been assumed to be Pareto or Poisson distributed in recent studies, e.g. Wong et al. (2017). Our Tactile Internet traffic models reveal which haptic packet interarrival time distributions best fit different types of teleoperation systems, while the assumption of Poisson traffic is found valid only for a very special case.

(ii) Second, we build on the recently proposed concept of so-called *FiWi enhanced LTE-A heterogeneous networks (HetNets)* (Beyranvand et al., 2017), which were shown to achieve the 5G and Tactile Internet key requirements of very low latency and ultrahigh reliability by unifying coverage-centric 4G mobile networks and capacity-centric FiWi broadband access networks based on data-centric Ethernet technologies. By means of probabilistic analysis and verifying simulations based on recent and comprehensive smartphone traces, Beyranvand et al. (2017) showed that an average end-to-end latency of 1 ms can be achieved for a wide range of traffic loads and that MUs can be provided with highly fault-tolerant FiWi connectivity for reliable low-latency fiber backhaul sharing and WiFi offloading. Note, however, that only conventional H2H communications was considered by Beyranvand et al. (2017). In this chapter, we investigate the coexistence of MUs and HOs/TORs and explore HITL-centric teleoperation techniques that achieve the aforementioned Tactile Internet target of 1 ms under different haptic traffic scenarios.

(iii) Third, for enhanced Tactile Internet reliability performance we present our proposed *edge sample forecast (ESF)* module, which is inserted at the edge of our communication network in close proximity to the HO, as shown in Figure 2.2. A concept, originally known as mobile edge computing, has been standardized by ETSI for 5G networks. Note that since September 2016, ETSI has dropped the "mobile" out of MEC and renamed it *multiaccess edge computing (MEC)* in order to broaden its applicability to HetNets, including WiFi and fixed access technologies (e.g. fiber) (Taleb et al., 2017). Our proposed ESF module leverages MEC servers with embedded artificial intelligence (AI) capabilities that are placed at the optical-wireless interface of FiWi enhanced LTE-A HetNets to compensate for delayed haptic samples in the feedback path by means of multiple-sample-ahead-of-time forecasting. In doing so, the response time of the HO can be kept small, resulting in a tighter togetherness with and thereby an improved safety in the remote TOR environment.

The remainder of the chapter is structured as follows: Section 2.2 sheds some light on the dichotomy between automation (i.e. replacement of capabilities) and augmentation (i.e. extension of capabilities) of the human through the Tactile Internet. Section 2.3 derives Tactile Internet traffic models from haptic traces by studying teleoperation as an example of an immersive Tactile Internet experience. Section 2.4 introduces the concept of low-latency FiWi enhanced LTE-A HetNets using advanced MEC capabilities. In Section 2.6, we propose our AI-based sample forecasting scheme to help an HO experience an immersive teleoperation experience. In Section 2.5, we develop our analytical framework to estimate end-to-end delay in teleoperation over FiWi enhanced mobile networks. Section 2.7 presents analytical latency results verified by haptic trace driven simulations. Finally, Section 2.8 concludes the chapter.

2.2 The Tactile Internet: Automation or Augmentation of the Human?

A new wave of technological change, the wave of computerization, automation, and robotization, will eliminate not only manual efforts but also more and more complex mental functions that until recently were carried out by humans, as already predicted by Wassily Leontief, the 1973 Nobel Laureate in Economics, (Leontief 1983). Leontief argued that the process by which progressive introduction of new computerized, automated, and robotized equipment can be expected to reduce the role of human labor is similar to the process by which the introduction of tractors and other machinery first reduced and then completely eliminated horses and other draft animals in agriculture. Even if horses were ready to accept smaller rations of oats or hay per working day, the process of their gradual elimination would slow down only temporarily. More and more efficient tractors would come along, and finally, unable to compete with the superior performance of machines, horses would lose their jobs.

Predictions that automation will make humans redundant have been made before. President John F. Kennedy declared that the major domestic challenge of the 1960s was to maintain full employment at a time when automation is replacing men. In 1964, a group of Nobel laureates, known as the Ad Hoc Committee on the Triple Revolution, alerted President Lyndon Johnson to the danger of a revolution triggered by the combination of the computer and the automated self-regulating machine, threatening to divide society into a skilled elite and an unskilled underclass. Such doomsday predictions have in common that they succumb to the so-called *lump of labor fallacy*,[3] assuming that there

---

[3] *The Economist*, "Automation and anxiety: Will smarter machines cause mass unemployment?," June 2016.

is only a finite amount of work and if some of it is automated then there is less for humans to do. Conversely, in a more recent study, the McKinsey Global Institute (2017) comes to the conclusion that automation instead could make us all more human (rather than redundant) by creating an opportunity to work more closely with technology, freeing up more time to make use of intrinsically human capabilities and innate human skills, which will be at a premium as machines take on ever more of the predictable activities of the workday. Like past technological changes, robotization can be a very good thing, relieving the workload of humans while helping overcome many challenges the world faces. Toward this end, however, *spreading ownership* of robots and machines across people whose work they replace will be crucial to mitigate the risk of dividing societies between the owners of the robots on one side and the workers, who compete with the robots/machines on the other, and reduce the risk of producing a new robot-age feudalism with unprecedented social inequality (Freeman, 2016).

At the nexus of computerization, automation, and robotization lies the emerging *Tactile Internet*, which will be centered around H2M/R communications by leveraging devices that enable tactile sensations (Maier et al., 2016). It holds promise of an Internet that will enable the delivery of skills in digital form globally (Dohler et al., 2017). The Tactile Internet is expected to cover a wide range of application fields, including remote healthcare, autonomous/assisted driving, entertainment, and industry automation. In most of these industry verticals, very low latency and ultrahigh reliability are key for realizing immersive applications such as robotic teleoperation. Note, however, that some use cases do not necessarily require mobility all the time and thus can be carried out in fixed broadband network environments. Hence, 5G cellular networks need to be fully converged networks, where different fixed and mobile access technologies can be flexibly selected while sharing core network functionalities, leading to latency and reliability improvements (Lema et al., 2017).

In this chapter, we leverage on our recently proposed concept of FiWi enhanced LTE-A HetNets, which were shown to achieve the 5G and Tactile Internet key requirements of very low latency on the order of 1 ms and ultra-high reliability by unifying coverage-centric 4G mobile networks and capacity-centric fiber-wireless broadband access networks based on low-cost, data-centric Ethernet NG-PON and Gigabit-class wireless local area network (WLAN) technologies (Beyranvand et al., 2017). While necessary, the design of reliable low-latency converged communication network infrastructures is not sufficient to realize the full potential of the Tactile Internet. In the following, we inquire into possibilities to further extend the capabilities of FiWi enhanced LTE-A HetNets, paying particular attention to the aforementioned dichotomy between automation and augmentation (i.e. extension of capabilities) of the human through the Tactile Internet.

Toward this end, we let us guide by the following contemporary as well as early-day Internet visionaries. In *The Inevitable*, Kevin Kelly argues that in terms of the Internet, nothing has happened yet (Kelly, 2016). He suggests that if you want a glimpse of what we humans do when the robots take our current jobs, look at experiences. Humans excel at creating and consuming experiences. This is no place for robots. Among other technological forces that will shape our future, Kelly highlights that *cognifying*, i.e. embedding AI into an existing process or inert thing, will be hundreds of times more disruptive to our lives than the transformations gained by industrialization. Ideally, according to Kelly, the additional intelligence should be not just cheap, but free, like the free commons of the web. In *Deep Thinking: Where Machine Intelligence Ends and Human Creativity Begins*, Garry Kasparov elaborates on the importance of superior process in human–machine collaboration, showing that *weak human + machine + better process* is superior to *strong human + machine + inferior process* (Kasparov 2017). Thus, a clever process beats superior knowledge and superior technology. His observation received interest by Google and other Silicon Valley companies and shifted the research focus from AI to intelligence amplification (IA) by using information technology as an *augmentation* tool to enhance human decisions (see, e.g. IBM's Watson) instead of replacing them with autonomous AI systems. According to Kasparov, this is not just user experience (UX), but entirely new ways of bringing human–machine coordination into diverse fields and creating the new tools we need in order to do so. Interestingly, this approach is fully in line with the original vision of early Internet pioneers. Back in 1962, Douglas C. Engelbart developed a detailed, though rudimentary, conceptual framework with process hierarchies for augmenting the human intellect by increasing via online assistance the capability of a man to derive solutions to complex problems that before seemed insoluble (Engelbart, 1962). Earlier, in 1960, Joseph C. R. Licklider envisioned *man-computer symbiosis*, a subclass of man–machine systems, to enable close interaction between man and computer in mutually beneficial cooperation (Licklider, 1960).

The Tactile Internet sets demanding requirements for future access networks in terms of latency, reliability, and also capacity (e.g. high data rates for video sensors). Wired access networks are partly meeting these requirements already, but wireless access networks are not yet designed to match these needs. A round-trip latency of 1–10 ms in conjunction with carrier-grade robustness and availability will enable the Tactile Internet for steering and control of real and virtual objects (Fettweis, 2014). Toward this end, the Tactile Internet will enable haptic communications and provide the medium for transporting touch and actuation in real time, i.e. the ability to exert haptic control through the Internet, in addition to nonhaptic control and data such that the end-user will not be able to tell the difference between controlling a system locally or from another location (Van Den Berg et al., 2017). For a comprehensive update on state-of-the-art haptic interfaces and telepresence systems we refer the interested reader to Prattichizzo et al. (2018).

The idea of remotely controlling machines via the Internet has been studied since the late 1990s. Luo et al. (1999) proposed the tele-control of a rapid prototyping machine (similar to a 3D printer) via the Internet for the purpose of automated telemanufacturing. Expanding on the idea to use sophisticated and expensive manufacturing facilities by several users around the world, the teleoperation issues related to the transmission of haptic information over the Internet were investigated in greater detail by Elhajj et al. (2001). Although the proposed control method was able to ensure stability, synchronization, and transparency in teleoperation, the reported round-trip time of packets transmitted between HO and TOR were above 250 ms, thus missing the Tactile Internet target of 1–10 ms.

Advanced cloud robotics and automation systems may be built by connecting them to the cloud in order to benefit from their networked operation via big data, cloud computing, collective robot learning, and crowdsourcing capabilities (Kehoe et al., 2015). Here, crowdsourcing is used to tap human skills for analyzing images and video, classification, learning, and error recovery, i.e. humans are used to enhance cloud robotics and automation systems. Note that unlike networked robots, where robots communicate with each other via local area networks, networked telerobots keep the *human in the loop*, where robots are operated remotely by humans via global networks, e.g. the Internet. According to Kehoe et al. (2015), an important open research challenge in cloud robotics is the development of new algorithms and methods to cope with time-varying latency, i.e. jitter. The high requirements of future haptic applications that allow full immersion demand ultra-reliable and low-latency communications (URLLC). For a comprehensive and up-to-date survey on methodologies and technologies for enabling URLLC infrastructures for haptic communications, we refer the interested reader to Antonakoglou et al. (2018) and the references therein.

Beside the design of URLLC infrastructures underlying the Tactile Internet, another key challenge little discussed in existent Tactile Internet surveys is how we can make sure that the potential of the Tactile Internet be unleashed for a race with (rather than against) machines. The overarching goal of the Tactile Internet should be the production of new goods and services by means of empowering rather than automating machines that complement humans rather than substitute for them. The Tactile Internet should amplify the differences between machines and humans. By building on the areas where machines are strong and humans are weak, the machines are more likely to complement humans rather than substitute for them. The value of human inputs will grow, not shrink, as the power of machines increases (Brynjolfsson and McAfee, 2014). In the future, coworking with robots will require human expertise in the coordination of the human–robot symbiosis. Clearly, while URLCC is necessary to meet the very low latency and ultrahigh reliability requirements of the Tactile Internet, it does not address the proper task assignment nor does it provide suitable mechanisms to orchestrate the mutually beneficial cooperation of humans and machines.

A promising approach toward achieving advanced human–machine coordination by means of a superior process for fluidly orchestrating human and machine coactivity may be found in the still young field of *human-agent-robot teamwork (HART)* research (Bradshaw et al., 2012). Unlike early automation research, HART goes beyond the singular focus on full autonomy (i.e. complete independence and self-sufficiency) and cooperative/collaborative autonomy among autonomous systems themselves, which aim at excluding humans as potential teammates for the design of human-out-of-the-loop solutions. In HART, the dynamic allocation of functions and tasks between humans and machines, which may vary over time or be unpredictable in different situations, plays a central role. In particular, with the rise of increasingly smarter machines, the historical humans-are-better-at (HABA)/machines-are-better-at (MABA) approach to decide which tasks are best performed by people and which by machines rather than working in concert has become obsolete. To provide a better understanding of the potential and limitations of current smart machines, T. H. Davenport and J. Kirby classified in *Only Humans Need Apply: Winners and Losers in the Age of Smart Machines* the capabilities of intelligent machines along two dimensions, namely, their ability to act and their ability to learn, as illustrated in Table 2.1 (Davenport and Kirby, 2016). In the vertical dimension, the ability to act involves four task levels, ranging from the most basic tasks (e.g. analyzing numbers) to performing digital tasks (done by agents) or even physical tasks (done by robots). In the horizontal dimension, the ability to learn escalates through four levels, spanning from human-support machines with no inherent intelligence to machines with context-awareness, learning, or even self-aware intelligence (to be elaborated on in technically greater detail in Chapter 3). The upper left of Table 2.1 consists of tasks doable by state-of-the-art machines. The lower and in particular far right of it is territory not yet conquered by machines.

According to Bradshaw et al. (2012), among other HART research challenges, the development of capabilities that enable autonomous systems not merely to do things *for* people but also to work together *with* people and other systems represents an important open issue in order to treat the human as a "member" of a team of intelligent actors rather than keep viewing him as a conventional "user."

In the following, after presenting an in-depth study of haptic traffic characterization and modeling, we introduce FiWi-enhanced LTE-A HetNets with AI-embedded MEC capabilities to achieve both low round-trip latency and low jitter. To showcase the achievable performance gains, we study the use case of HART-centric teleoperation via simulation based on haptic Tactile Internet traffic traces.

Table 2.1 Classification of intelligent machines along two dimensions: Ability to act (vertical) and ability to learn (horizontal).

Task type	Human support	Repetitive task automation	Context-awareness and learning	Self-aware intelligence
Analyze numbers	Business intelligence, data visualization, hypothesis-driven analytics	Operational analytics, scoring, model management	Machine learning, neural networks	Not yet
Digest words, images	Character and speech recognition	Image recognition, machine vision	Watson, natural language processing	Not yet
Perform digital tasks (Admin and decisions)	Business process management	Rules engines, robotic process automation	Not yet	Not yet
Perform physical tasks	Remote operation	Industrial robotics, collaborative robotics	Fully autonomous robots, vehicles	Not yet

2.3 Haptic Traffic Characterization

An interesting example of a Tactile Internet experience that allows for remote immersion is the HART-centric use case of teleoperation based on *haptic communications*. As mentioned earlier, the Tactile Internet envisions the real-time transmission of haptic information for the remote control of physical and/or virtual objects through the Internet (Antonakoglou et al., 2018). Recall from Section 2.1 that in a typical bilateral teleoperation system, the HO interfaces with the communication network (to be described in greater detail in Section 2.4) via the HSI device, which is used to display haptic interaction with the remote TOR to the HO (see Figure 2.4). A perceptual deadband-based (i.e. zero output if changes in consecutive samples are minimal) data reduction may be deployed as a lossy compression mechanism by exploiting the fact that human end-users are not able to discriminate arbitrarily small differences in haptic stimuli. The human perception of haptics can be exploited to reduce the haptic packet rate. Specifically, the well-known Weber's law determines the just noticeable difference (JND), i.e. the minimum change in the magnitude of a stimulus that can be detected by humans Weber (1978). Weber's law gives rise to the so-called *deadband coding* technique, whereby a haptic sample is transmitted only if its change with respect to the previously transmitted haptic sample exceeds a given deadband parameter $d \geq 0$ (given in percent) (Steinbach et al., 2012).

In the following, we take a closer look at the specific characteristics of Tactile Internet traffic by studying the use case of teleoperation. Specifically, we study two sets of haptic traces obtained from teleoperation experiments involving TORs

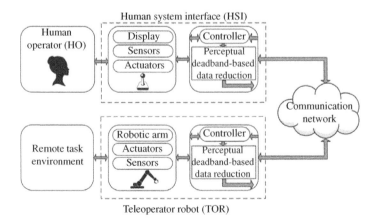

Figure 2.4 Teleoperation system based on bidirectional haptic communications between HO and TOR in a remote task environment. Source: Maier and Ebrahimzadeh (2019). © 2019 IEEE.

with different DoF. The two considered teleoperation experiments involve TORs with 1 and 6 DoF. Furthermore, our haptic traces comprise measurements with different values of deadband parameter d.

2.3.1 Teleoperation Experiments

2.3.1.1 6-DoF Teleoperation without Deadband Coding

The first set of our traces for a haptics-enabled telesurgery system were provided by Meli et al. (2017) from the Centre National de la Recherche Scientifique (CNRS) at IRISA, Rennes, France. Note that telesurgery represents a well-known type of teleoperation in the healthcare sector. The system consists of a 6-DoF haptic interface at the HO side, a 6-DoF manipulator, and a six-axes force/torque sensor at the TOR side. Update samples containing the position and orientation signals from the HO are transmitted at every refresh time instant. Similarly, the HO receives force–torque feedback samples from the remote TOR. The local HO and remote TOR environment were put back-to-back during the experiments, i.e. there were no communication-induced artifacts such as latency. Note that deadband coding was not applied in this 6-DoF telesurgery experiment, i.e. $d = 0$.

2.3.1.2 1-DoF Teleoperation with Deadband Coding

The second set of haptic traces were obtained from the 1-DoF teleoperation experiments at the Technical University of Munich, Germany (Xu et al., 2016). Two Phantom Omni[4] devices were used as master (i.e. HO) and slave (i.e. TOR) devices to create a 1-DoF bilateral teleoperation scenario. The communication channel between HO and TOR was emulated by using a variable queuing system to generate constant or time-varying delays. The velocity signal at the HO side was sampled before being transmitted to the TOR, which in turn fed the force signal back to the HOR. The experiments were run with different deadband values set to $d \in \{0\%, 5\%, 10\%, 15\%, 20\%\}$ in both the command and feedback paths.

2.3.1.3 Packetization

Typically, haptic samples are packetized and transmitted immediately once new sensor readings are available to help minimize the end-to-end delay, implying a real-time transport protocol (RTP), user datagram protocol (UDP), and Internet Protocol (IP) header of 12, 8, and 20 bytes, respectively Steinbach et al. (2012). Additionally, for each DoF, the haptic sample of the aforementioned experimental sensor readings comprises 8 bytes. Note that N_{DoF} haptic samples are encapsulated

---

4 Phantom Omni is a widely used HSI device that enables HOs to interact with and manipulate objects by adding 3D navigation to a broad range of applications, e.g. games, entertainment, visualization, among others.

into one RTP/UDP/IP packet, where N_{DoF} denotes the number of DoF in either experiment (i.e. 6 or 1 in our case). Thus, the packet size is equal to $40 + 8 \cdot N_{\text{DoF}}$ bytes.

2.3.2 Packet Interarrival Times

We begin by investigating the packet interarrival times of both teleoperation traces. For a given deadband parameter d, let $\lambda^c(d_c)$ and $\lambda^f(d_f)$ denote the mean packet rate at which packets arrive at the MAC layer of the wireless interface in the command path and feedback path, respectively. In the following, we discuss both teleoperation traces separately, first without deadband coding ($d = 0$) and then with deadband coding ($d > 0$).

Note that in our 1-DoF teleoperation traces without deadband coding, packet interarrival times are deterministic with a constant packet arrival rate of $\lambda^c(d_c)|_{d_c=0} = \lambda^f(d_f)|_{d_f=0} = 1000$ packets/s in both command and feedback paths due to the fixed haptic sampling rate of 1 kHz. Conversely, in our 6-DoF teleoperation traces without any deadband coding, haptic samples are immediately packetized and transmitted at varying (i.e. nondeterministic) refresh time instants. In the following, we examine the command and feedback paths of our 6-DoF teleoperation traces separately and try to find the best fitting distribution for the respective packet interarrival times.

First, let us focus on the position and orientation samples in the command path (from HO to TOR), which are measured as a triplet and quaternion (i.e. quadruple) at each refresh time instant, respectively. Let **COMD**$_i$ denote the resultant position-orientation sample i, which is transmitted as packet in the command path at time instant $T_i^{(c)}$. Thus, the corresponding packet interarrival times $I_i^{(c)} = T_i^{(c)} - T_{i-1}^{(c)}$, $i = 2, 3, \ldots$, represent realizations of the random variable $I^{(c)}$. Figure 2.5a depicts the histogram of the packet interarrival times $I^{(c)}$ in the command path obtained from the 6-DoF teleoperation traces, with the most frequent packet interarrival time expectedly being centered at 1 ms due to the default haptic sampling rate of 1 kHz.

The histogram of the packet interarrival times $I^{(f)}$ in the feedback path (from TOR to HO) is shown in Figure 2.5b. Interestingly, the feedback path differs from the command path in that it exhibits two peaks at approximately 0.75 ms and another one at 1.25 ms. Upon examining the force/torque traces stemming from the TOR side, we found that the two peaks exist because the force and torque sensors of the TOR operate at two slightly different sampling rates above and below 1 kHz.

In an effort to find a probability distribution function (PDF) that best fits the experimental packet interarrival times in Figure 2.5, we considered a variety of well-known distributions. Our preliminary evaluations narrowed our choice

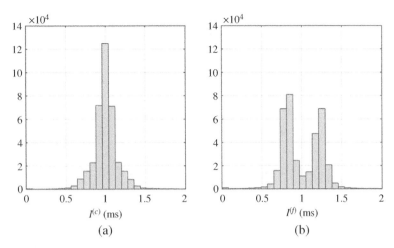

Figure 2.5 Histogram of experimental 6-DoF teleoperation packet interarrival times: (a) command path and (b) feedback path. Source: Maier and Ebrahimzadeh (2019). © 2019 IEEE.

down to three candidate PDFs, namely, exponential, generalized Pareto (GP), and gamma distributions. Our method of selecting the best fitting PDF comprised the following three steps. First, we used the maximum likelihood estimation (MLE) method to estimate the parameters of each PDF. Second, the estimates of the first step were verified by computing the complementary cumulative distribution function (CCDF) $F_{I^{(c)}}(\zeta) = P(I^{(c)} > \zeta)$. Third, to compare the goodness-of-fit among the three PDFs under consideration, we used the maximum difference D^* between the fitted and experimental CCDFs, which is given by

$$D^* = \sup_{\zeta} |\hat{F}_{I^{(c)}}(\zeta) - F_{I^{(c)}}(\zeta)| \tag{2.1}$$

whereby $\hat{F}_{I^{(c)}}(\zeta)$ denotes the experimental CCDF. The estimated parameters as well as the calculated D^* of the fitting PDFs are listed in Table 2.2, where we observe that the gamma distribution matches the experimental data reasonably well, as opposed to the exponential and GP distributions. Next, we proceed by fitting the best PDF to the 6-DoF experimental packet interarrival times in the feedback path. Similar to the command path, we observe from Table 2.2 that in the feedback path, the gamma distribution again fits the experimental data best.

Figure 2.6a shows the CCDF of the three fitted PDFs and experimental 6-DoF teleoperation packet interarrival times in the command path. We observe from the figure that the gamma distribution matches the experimental data reasonably well, as opposed to the exponential and GP distributions. Similar to the command path, we observe from Figure 2.6b and Table 2.2 that for the CCDF in the feedback path, $F_{I^{(f)}}(\zeta) = P(I^{(f)} > \zeta)$, the gamma distribution again best fits the experimental data.

Table 2.2 Summary of the estimated parameters of fitted PDFs using MLE method.

		Exponential $f_l(x) = \dfrac{1}{\mu} e^{-\frac{x}{\mu}}$, $x \ge 0$		Generalized Pareto $f_l(x) = \dfrac{1}{\sigma}\left(1 + k\dfrac{x-\theta}{\sigma}\right)^{-1-\frac{1}{k}}$, $x \ge \theta$				Gamma $f_l(x) = \dfrac{r^a}{\Gamma(a)} x^a \dfrac{e^{-rx}}{x}$, $x \ge 0$		
	$d\,(\%)$	μ	D^*	k	σ	θ	D^*	a	r	D^*
6-DoF (Command path)	0	0.0010	0.47	−0.065	1.0×10^{-3}	3.7×10^{-6}	0.46	27	27620	**0.14**
	0.05	0.0028	0.23	0.12	2.4×10^{-3}	4.7×10^{-6}	**0.17**	1.51	540	0.19
	0.10	0.0047	0.16	0.16	3.8×10^{-3}	9.6×10^{-6}	**0.14**	1.29	270	0.19
	0.20	0.0087	0.18	0.19	6.6×10^{-3}	7.15×10^{-6}	**0.13**	0.65	27	0.15
6-DoF (Feedback path)	0	0.001	0.45	−0.064	1.0×10^{-3}	2.6×10^{-6}	0.45	12	12166	**0.10**
	5	0.0012	0.41	−0.02	1.2×10^{-3}	2.8×10^{-6}	0.40	4.85	4121	**0.19**
	10	0.0014	0.36	0.05	1.3×10^{-3}	4.1×10^{-6}	0.37	2.56	1877	**0.23**
	20	0.0017	**0.21**	0.13	1.3×10^{-3}	2.8×10^{-6}	0.27	1.54	931	0.31
1-DoF (Command path)	5	0.0022	0.36	0.63	5.7×10^{-4}	7.3×10^{-4}	**0.34**	1.46	663	0.38
	10	0.0027	0.38	0.81	4.9×10^{-4}	7.5×10^{-4}	**0.34**	1.04	386	0.39
	15	0.0038	0.41	0.88	7.8×10^{-4}	6.4×10^{-4}	**0.32**	0.79	208	0.36
1-DoF (Feedback path)	5	0.0075	0.16	0.46	3.7×10^{-3}	5.8×10^{-4}	**0.10**	0.91	120	0.14
	10	0.0024	0.20	0.57	11.5×10^{-4}	3.4×10^{-4}	**0.05**	0.69	28	0.13
	15	0.0036	0.12	0.32	24.2×10^{-3}	11.2×10^{-4}	**0.04**	0.91	25	0.10

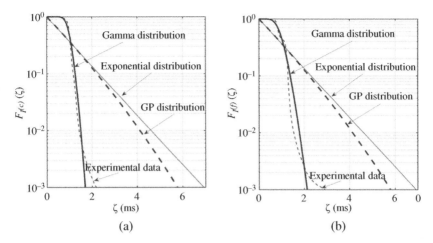

Figure 2.6 Complementary cumulative distribution function (CCDF) of fitted probability distribution functions (PDFs) and experimental 6-DoF teleoperation packet interarrival times: (a) command path and (b) feedback path. Source: Maier and Ebrahimzadeh (2019). © 2019 IEEE.

Next, we study the 1-DoF teleoperation experiment, which included deadband coding unlike its 6-DoF counterpart. For fair comparison of the two sets of haptic traces, we postprocessed the original 6-DoF traces and applied deadband coding for a variety of different deadband parameter values in the command path (d_c) and feedback path (d_f), as explained in the following. To begin with, we model the position signal with a 3D vector-valued function of time denoted by $\mathbf{p}(t)$. Further, let $\mathbf{o}(t)$ denote the orientation signal, which is modeled by a quaternion.[5] Similar to the position signal in the command path, we model the force and torque signals in the feedback path by 3D vector-valued function $\mathbf{f}(t)$ and $\mathbf{t}(t)$, respectively. We apply the deadband coding as follows. In the command path of 6-DoF teleoperation, with deadband coding, a position-orientation sample $\mathbf{comd}(t)$ is transmitted only if the proportional change with respect to the previously transmitted sample $\mathbf{COMD}_{i-1}$ exceeds a given d_c, i.e. $\mathbf{COMD}_i = \mathbf{comd}(t)$ only if

$$\max\{\Delta_p, \Delta_o\} > d_c$$

where $\Delta_p = \frac{\|\mathbf{p}(t) - \mathbf{p}_{i-1}\|}{\|\mathbf{p}_{i-1}\|}$ and $\Delta_o = \max\{\Delta_{\hat{v}}, \Delta_{\hat{\theta}}\}$, whereby $\Delta_{\hat{v}} = \frac{\|\hat{v}(t) - \hat{v}_{i-1}\|}{\|\hat{v}_{i-1}\|}$ and $\Delta_{\hat{\theta}} = \frac{\|\hat{\theta}(t) - \hat{\theta}_{i-1}\|}{\|\hat{\theta}_{i-1}\|}$. Note that $\|\cdot\|$ denotes the Euclidean norm function. In the feedback

---

5 Quaternion representation of orientation is characterized by $\hat{v} = (\hat{v}_x, \hat{v}_y, \hat{v}_z)$ and $\hat{\theta}$, where $\hat{\theta}$ is the angle of rotation and $\hat{v}$ is the unit vector about which rotation is performed, i.e. the axis of rotation.

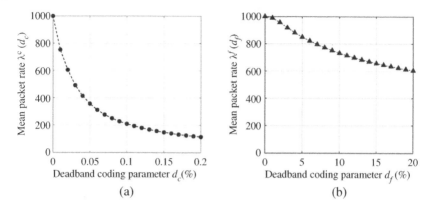

Figure 2.7 Mean packet rate (in packets/s) vs. d for 6-DoF teleoperation: (a) command path and (b) feedback path. Source: Based on Maier and Ebrahimzadeh (2019). © 2019 IEEE.

path of 6-DoF teleoperation, an update force–torque sample is transmitted, only if $\max\{\Delta_f, \Delta_t\} > d_f$, where $\Delta_f = \frac{\|\mathbf{f}(t)-\mathbf{f}_{i-1}\|}{\|\mathbf{f}_{i-1}\|}$ and $\Delta_t = \frac{\|\mathbf{t}(t)-\mathbf{t}_{i-1}\|}{\|\mathbf{t}_{i-1}\|}$.

Figure 2.7 illustrates the beneficial impact of deadband coding on reducing the haptic packet rate in the feedback path and in particular the command path. More specifically, note that in the command path, a deadband parameter of only $d_c = 0.02\%$ decreases the mean packet rate $\lambda^c(d_c)$ to roughly 600 packets/s, translating into a haptic packet rate reduction of 39.5% compared to the case without deadband coding (i.e. $d_c = 0$). As shown in Figure 2.7a, $\lambda^c(d_c)$ further decreases for increasing d_c and levels off for $d_c > 0.1\%$. We observe from Figure 2.7b that deadband coding is less effective in the feedback path, where a deadband parameter of as high as $d_f = 20\%$ (compared to $d_c = 0.02\%$ above) is needed to reduce the mean packet rate $\lambda^f(d_f)$ to roughly 600 packets/s.

We again determined the best fitting PDFs for the packet interarrival times with the different deadband parameter values by following the same approach as described above. Table 2.2 comprehensively summarizes our findings on the different best fitting packet interarrival time distributions for command and feedback paths with and without deadband coding in both teleoperation scenarios under consideration. Note that in Table 2.2, the goodness-of-fit of the best fitting PDF for each teleoperation scenario is shown in bold.

For completeness, Figure 2.8 comprehensively summarizes our findings on the different best fitting packet interarrival time distributions for command and feedback paths with and without deadband coding in both teleoperation scenarios. We observe that in general, command and feedback paths can be jointly modeled by the GP, gamma, or deterministic packet interarrival time distribution, depending on the given value of deadband parameters d_c and d_f, as shown in

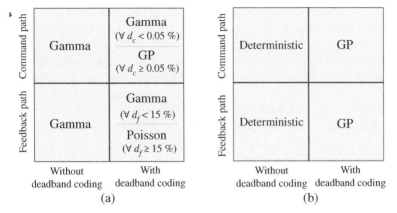

Figure 2.8 Summary of best fitting packet interarrival time distributions for command and feedback paths with and without deadband coding: (a) 6-DoF teleoperation and (b) 1-DoF teleoperation. Source: Maier and Ebrahimzadeh (2019). © 2019 IEEE.

Figure 2.8. Importantly, our haptic trace analysis indicates that the assumption made in recent studies that Tactile Internet traffic is Pareto distributed and is not valid for the analyzed traffic. Furthermore, the assumption of Poisson traffic (e.g. (Wong et al., 2017)) with exponentially distributed packet interarrival times was found valid only for 6-DoF teleoperation in the feedback path with deadband parameter values of $d_f \geq 15\%$. We note that our trace analysis provides important yet preliminary insights into the statistics of Tactile Internet traffic. Clearly, a more systematic approach looking at additional haptic traces of different types of teleoperation experiments will be instrumental in accurately validating the packet interarrival time distributions reported above.

2.3.3 Sample Autocorrelation

After modeling haptic traffic arrival, we take a closer look at our available traces in the feedback path to identify possible correlation patterns in haptic samples. Such correlation patterns can be useful in developing sample forecasting techniques, leveraging AI capabilities at the network edge to compensate for delayed feedback samples by making accurate forecasts (Maier and Ebrahimzadeh, 2019), to be discussed in technically great detail later on. In the following, we are going to answer the following questions: (i) how deep are the feedback samples correlated with their own lagged samples? and (ii) what is the impact of deadband coding on the autocorrelation of the feedback samples?

To answer these questions, we devise the autocorrelation function. We note, however, that haptic packets transmitted in a typical teleoperation system contain the samples taken from continuous signals, which are either 1D (i.e. force signal in 1-DoF teleoperation) or 3D (i.e. force/torque signal in 6-DoF teleoperation) vector-valued functions of time. Unlike 1D signals, estimating the autocorrelation function of a multidimensional vector-valued function is not straightforward. We thus present our method of estimating autocorrelation function of a given multidimensional discrete vector-valued function. For the sake of argument, let us consider $\mathbf{x}(t)$ as a 3D vector-valued function evaluated at time t, which is characterized by $\mathbf{x}(t) = x_1(t)\mathbf{i} + x_2(t)\mathbf{j} + x_3(t)\mathbf{k}$, where $x_1(t)$, $x_2(t)$, and $x_3(t)$ are the corresponding $x-$, $y-$, and $z-$coordinates of $\mathbf{x}(t)$, respectively. Note that $\mathbf{i}$, $\mathbf{j}$, and $\mathbf{k}$ are unit vectors representing the axes of the Cartesian coordinate system. We estimate the sample mean $\overline{\mathbf{x}} \in \mathbb{R}^3$ of a given vector-valued function $\mathbf{x}(t)$ by $\frac{1}{N_s}\sum_{i=1}^{N_s}\mathbf{x}(i\overline{T}_s)$, where N_s and $\overline{T}_s$ denote the total number of samples and inter-sample time, respectively. We then estimate the sample variance $\sigma_{\mathbf{x}}^2 \in \mathbb{R}^+ \cup \{0\}$ by $\frac{1}{N_s-1}\sum_{i=1}^{N_s}\| \mathbf{x}(i\overline{T}_s) - \overline{\mathbf{x}} \|^2$, which can be generalized to estimate the autocorrelation function $\hat{R}_{\mathbf{x}}(h)$ by $C(h)/\sigma_{\mathbf{x}}^2$, where $C(h)$, $\forall h \ll N_s$, is given by

$$C(h) = \frac{1}{N_s-1}\sum_{i=1}^{N_s-h} \ll (\mathbf{x}(i\overline{T}_s) - \overline{\mathbf{x}}) \cdot (\mathbf{x}((i+h)\overline{T}_s) - \overline{\mathbf{x}}) \gg \qquad (2.2)$$

where $\ll \cdot \gg$ denotes the inner product.[6]

Figure 2.9 depicts the autocorrelation function of the force/torque feedback samples of both available sets of traces for different teleoperation scenarios with and without deadband coding. To cope with irregular sampling intervals, which occur after performing deadband coding, we have used a zero-order hold interpolator at the rate of 1 kHz. We observe that the force/torque samples represent a quite deep correlation with their own lagged samples. Let correlation depth h_α^* denote the maximum time lag such that, for $h < h_\alpha^*$, force/torque sample autocorrelation $\hat{R}(h)$ is greater than $\alpha\%$. Note that deadband coding, in general, decreases the autocorrelation of feedback samples for a given time lag h, thus decreasing the correlation depth, see Figure 2.9a–c. For instance, the force feedback signal in 6-DoF teleoperation without deadband coding exhibits a correlation depth $h_{90\%}^*$ of 202. Deadband coding, in turn, reduces the correlation depth to 142, 116, and 98 for $d = 5\%$, $d = 10\%$, and $d = 15\%$, respectively. Further, we find that the torque samples show a slightly higher autocorrelation compared with that of the force samples in 6-DoF teleoperation. Also note that the 1-DoF force samples with deadband coding are associated with less autocorrelation, compared to the 6-DoF force samples. This is mainly due to the fact that in 6-DoF teleoperation, 3D force samples are transmitted, as opposed to 1-DoF

---

6 Note that our defined autocorrelation function is based on the notion of the inner product of vectors $\mathbf{x_1}$ and $\mathbf{x_2}$, which is given by $\| \mathbf{x_1} \| \cdot \| \mathbf{x_2} \| \cdot \cos\theta$, where θ is the angle between $\mathbf{x_1}$ and $\mathbf{x_2}$ in a multidimensional space. As θ deviates from zero, the two vectors are less correlated and vice versa.

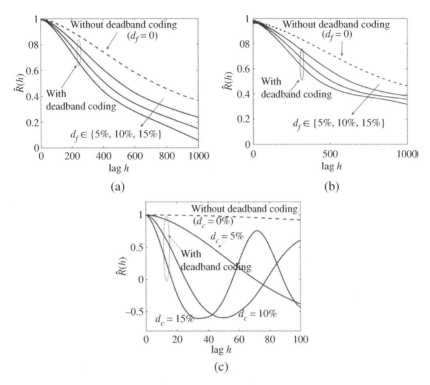

Figure 2.9 Estimation of the autocorrelation of the haptic samples in the feedback path of 1- and 6-DoF teleoperation: (a) force samples in 6-DoF teleoperation, (b) torque samples in 6-DoF teleoperation, and (c) force samples in 1-DoF teleoperation. Source: Based on Maier and Ebrahimzadeh (2019).

teleoperation, where only one dimensional force samples are fed back, thus being more susceptible to deadband coding.

2.4 FiWi Access Networks: Revisited for Clouds and Cloudlets

DBA is one of the contributing factors to latency in FiWi access networks, thus it is important to understand how the edge architecture with its associated MAC protocol(s) affects the overall DBA strategy. Decentralized DBA is preferred in order to eliminate the delays inherent with a centralized scheme. This is especially critical for the latency-driven applications of the Tactile Internet.

Recently, Hossain et al. (2017) provided a preliminary study of the shortcomings of centralized optical line terminal (OLT)-based DBA algorithms in FiWi access networks consisting of a conventional time division multiplexing (TDM) PON with a passive power splitter at the remote node and BSs connected to ONUs. Each BS was assumed to wirelessly exchange its queue information with other BSs in an attempt to build and maintain global knowledge among all BSs and thereby facilitate the use of a distributed DBA algorithm among all ONU-BSs. In doing so, the OLT was exempt from the upstream bandwidth allocation process, thus avoiding the detrimental impact of PON propagation delays in traditional centralized DBA algorithms. Despite the reported queuing delay performance improvements over the well-known centralized DBA algorithm interleaved polling with adaptive cycle time (IPACT), the wireless exchange of periodic control messages among all BSs may not be scalable in the wireless front-end. Furthermore, subscribers may access the wireless medium without the network assistance of BSs in a truly distributed manner, as explained in the two architectures described next.

2.4.1 FiWi: EPON and WLAN

Although a few FiWi architectural studies exist on the integration of PON with long-term evolution (LTE) or WiMAX wireless front-end networks, the vast majority of studies consider FiWi access networks consisting of a conventional IEEE 802.3ah Ethernet passive optical network (EPON) fiber backhaul and an IEEE 802.11b/g/n/s WLAN mesh front-end, which may be further upgraded by leveraging NG-PONs, notably 10+ Gb/s TDM/WDM PONs, and Gigabit-class IEEE 802.11ac very high throughput (VHT) WLAN technologies (Aurzada et al., 2014). Thus, most FiWi access networks rely on low-cost data-centric optical fiber Ethernet (EPON) and wireless Ethernet (WLAN) technologies, which provide a couple of important benefits. First, economic considerations are expected to play an even more critical role in 5G networks than in the previous four generations. Second, today's service providers have to cope with an unprecedented growth of mobile data traffic worldwide. Complementing 4G LTE-A HetNets with already widely deployed WiFi access points represents a key aspect of the strategy of today's operators to offload mobile data traffic from their cellular networks, a technique known as *WiFi offloading*. FiWi access networks with a WLAN-based front-end represent a promising approach to realize WiFi offloading in a cost-efficient manner.

Now, it is important to understand that, unlike LTE, WLANs use a *distributed* MAC protocol for arbitrating access to the wireless medium among stations. Specifically, the so-called distributed coordination function (DCF) typically deployed in WLANs may suffer from a seriously deteriorated throughput performance due to the propagation delay of the fiber backhaul. To see this, note that in WLANs a wireless source station starts a timer after each frame transmission and waits for the acknowledgment (ACK) from the wireless destination station. If the source station does not receive the ACK before the ACK timeout, it will resend the frame for a certain number of retransmission attempts. Clearly, one solution to compensate for the fiber propagation delay is to increase the ACK timeout. Note, however, that in DCF the ACK timeout must not exceed the DCFinterframe space (DIFS), which prevents other stations from accessing the wireless medium, thus avoiding collision with the ACK frame (in IEEE 802.11 WLAN specifications DIFS is set to 50 μs). Due to the ACK timeout, backhaul fiber can be deployed in WLAN-based FiWi networks only up to a maximum length. For instance, in a standard IEEE 802.11b WLAN network with a default ACK timeout value of 20 μs, the backhaul fiber length must be less than 1948 meters to ensure the proper operation of DCF.

Clearly, the aforementioned limitations of WLAN-based FiWi access networks can be avoided by controlling access to the optical fiber and wireless media separately from each other, giving rise to so-called "radio-and-fiber" (R&F) networks (Maier et al., 2008). R&F-based FiWi access networks may deploy a number of enabling optical and wireless technologies, including tunable lasers and receivers, colorless ONUs, as well as burst-mode laser drivers and receivers. In RoF networks, optical fiber is used as an analog or digital transmission medium between a central station and one or more remote antenna units with the central station in charge of controlling access to both optical and wireless media. In contrast, in R&F networks, access to the optical and wireless media is controlled separately by using in general two different MAC protocols in the optical and wireless media, with protocol translation taking place at their optical-wireless interface. As a consequence, wireless MAC frames do not have to travel along the backhaul fiber to be processed at any central control station, but simply traverse their associated access point and remain in the WLAN. Access control is done locally inside the WLAN in a fully decentralized fashion, thus avoiding the negative impact of fiber propagation delay. Note that in doing so, WLAN-based FiWi access networks of extended coverage can be built without imposing stringent limits on the length of the fiber backhaul. Recall that this holds only for distributed MAC protocols such as DCF, but not for MAC protocols that deploy centralized polling and scheduling such as EPON and LTE. Thus, in a typical R&F-based FiWi access network consisting of a cascaded EPON backhaul and WLAN front-end for WiFi offloading, the end-to-end coordination of both fiber and wireless network resources may be done by a co-DBA algorithm that uses the centralized IEEE 802.3ah multipoint control protocol (MPCP) for EPON and the decentralized DCF for WiFi, with MAC protocol translation taking place at the optical-wireless interface. Note that the decentralized nature of WLAN's access protocol DCF is instrumental in realizing low-latency FiWi enhanced LTE-A HetNets, as explained in more detail in the subsequent section.

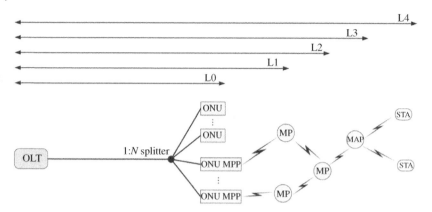

Figure 2.10 Hierarchical frame aggregation involving different aggregation layers (L0-L4) across Ethernet PON (EPON) backhaul and wireless local area network (WLAN) mesh front-end. Source: Adapted from Ghazisaidi and Maier (2011).

Next, let us consider an illustrative example to better understand the operation of co-DBA in FiWi access networks. A major MAC enhancement technique of next-generation WLANs is frame aggregation, which groups multiple wireless MAC frames into a single aggregate MAC protocol or service data unit for wireless transmission. In Ghazisaidi and Maier (2011), the benefits of co-DBA were demonstrated by extending advanced frame aggregation techniques to EPON and their integrated operation across both optical and wireless segments. The proposed hierarchical frame aggregation techniques involve different aggregation layer, ranging from hop-by-hop to end-to-end aggregation of traffic between the OLT and wireless stations, and help improve the throughput-delay performance of R&F-based FiWi access networks for voice, video, and data traffic. For illustration, Figure 2.10 depicts a FiWi access network consisting of a cascaded EPON backhaul and a WLAN-based mesh front-end. The wireless mesh front-end comprises mesh portal points (MPPs) collocated with ONUs, intermediate mesh points (MPs), and mesh access points (MAPs), each serving associated wireless stations (STAs). The five different aggregation layers (L0–L4), shown in the upper part of Figure 2.10, illustrate the possible operation modes of the co-DBA algorithm. Specifically, L0 applies frame aggregation only for traffic between the OLT and ONU-MPPs (as well as conventional ONUs without wireless extension), i.e. frame aggregation is used in the optical network segment separately from the wireless network segment. The remaining four aggregation layers (L1–L4) apply frame aggregation across the optical-wireless interface, thereby allowing for joint frame aggregation in both optical and wireless network segments, ranging from wireless single-hop MPs to all wireless multi-hop MPs, MAPs, and STAs in an end-to-end fashion.

2.4.2   C-RAN: Cloud vs. Cloudlet

In Maier and Rimal (2015), we studied FiWi access networks in the context of both conventional clouds and emerging cloudlets, paying particular attention to the difference between R&F and traditional RoF networks. RoF networks were used in, for example, China Mobile's C-RAN, which relies on a centralized cloud infrastructure and moves BBUs away from RRHs, intentionally rendering the RRHs as simple as possible without any processing and storage capabilities. Conversely, beside MAC protocol translation, the distributed processing and storage capabilities inherently built into R&F networks may be exploited for realizing a number of additional network functions. Therefore, Maier and Rimal (2015), we argued that R&F-based FiWi access networks may become the solution of choice in light of the aforementioned trends of 5G mobile networks toward decentralization based on cloudlets and MEC. For completeness, however, we note that R&F and RoF technologies may be also used jointly for providing multitier cloud computing services, which accommodate both central cloud (e.g. C-RAN) and decentralized edge computing services over the same network infrastructure. For further details on multitier cloud computing in FiWi-enhanced mobile networks, the interested reader is referred to Rimal et al. (2017a,b, 2018).

2.4.3   Low-Latency FiWi Enhanced LTE-A HetNets

Recall from Section 2.1 that FiWi access networks provide a promising approach to offload mobile data from cellular networks by means of WiFi offloading. Recent backhaul-aware 4G studies have begun to investigate the performance-limiting impact of backhaul links in small-cell networks, though most of them did not take fiber link failures into account and assumed the reliability of the backhaul to be ideal (i.e. offering an availability of ~100%).

To meet the URLLC requirements of 5G networks, Beyranvand et al. (2017) recently explored the performance gains obtained from enhancing coverage-centric 4G LTE-A HetNets with capacity-centric FiWi access networks based on low-cost, data-centric Ethernet NG-PON and Gigabit-class WLAN technologies. Clearly, by unifying LTE-A HetNets and FiWi access networks, low-cost high-speed mobile data offloading is achievable via high-capacity fiber backhaul (e.g. IEEE 802.3av 10G-EPON) and Gigabit-class WLAN that has been able to consistently provide data rates 100 times higher than cellular networks (Fettweis and Alamouti, 2014), thus helping reach the envisioned thousandfold gains in area capacity and 10 Gb/s peak data rates of 5G. In the following, we extend the concept of FiWi-enhanced LTE-A HetNets in order to enable both local and nonlocal teleoperation by exploiting AI-enhanced MEC capabilities. Note that neither teleoperation nor edge intelligence were addressed by Beyranvand et al. (2017).

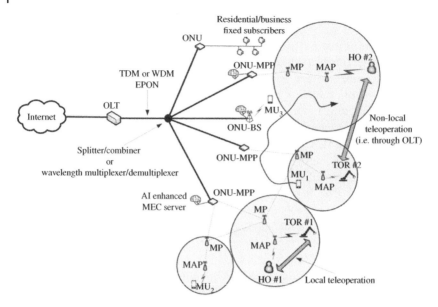

Figure 2.11 Local and non-local teleoperation in fiber-wireless (FiWi) enhanced-LTE-Advanced (LTE-A) Heterogenous Networks (HetNets) with artificial intelligence (AI)-enhanced multi-access edge computing (MEC) capabilities. Source: Maier and Ebrahimzadeh (2019). © 2019 IEEE.

Figure 2.11 depicts the generic network architecture of FiWi-enhanced LTE-A HetNets. The fiber backhaul consists of a TDM/WDM IEEE 802.3ah/av 1/10 Gb/s EPON with a typical fiber range of 20 km between the central OLT and remote ONUs. The EPON may comprise multiple stages, each stage separated by a wavelength-broadcasting splitter/combiner or wavelength multiplexer/demultiplexer. There are three different subsets of ONUs. An ONU may either serve fixed (wired) subscribers. Alternatively, an ONU may connect to either a cellular network BS or an IEEE 802.11n/ac/s WLAN MPP, giving rise to a collocated ONU-BS or ONU-MPP, respectively. Depending on current location and trajectory, a MU may communicate through the cellular network and/or WLAN mesh front-end, which consists of ONU-MPPs, intermediate MPs, and MAPs. Note that connecting these three different sets of ONUs via a common shared EPON fiber backhaul infrastructure helps achieve the important goal of fixed-mobile convergence gain of today's network operator strategy.

Beyranvand et al. (2017) proposed various advanced fiber-lean backhaul redundancy strategies (not shown in Figure 2.11), which can be used to realize a local Fx-FH with direct inter-ONU communication. Specifically, the following three strategies can be considered: (i) interconnection fiber links between pairs of neighboring ONUs, (ii) small-scale fiber protection rings among multiple nearby ONUs, and (iii) wireless bypassing of backhaul fiber faults via the WLAN front-end. The results showed that the localized protection techniques proposed in Beyranvand et al. (2017) are instrumental in providing fixed wired and MUs with highly fault-tolerant FiWi connectivity. Fx-FH solutions also help reduce latency by forming local clusters of ONUs as well as ONU-MPPs, thereby increasing the diversity of network connections. The analytical results verified by recent comprehensive smartphone traces showed that the presented interconnection fiber, fiber protection ring, and wireless protection techniques are able to keep the FiWi connectivity probability of MUs essentially flat for a wide range of EPON fiber-link failure probabilities while decreasing the average end-to-end delay to 1 ms for a wide range of traffic loads.

To better understand the reason behind the low delay performance of FiWi-enhanced LTE-A HetNets, we note that LTE systems themselves cannot guarantee low latency due to the fact that the transmission time interval is 1 ms. Thus, both uplink and downlink transmissions take at least 1 ms, translating into an end-to-end delay being lower bounded by 2 ms. In real-world deployment scenarios, the latency in LTE networks may increase by an order of magnitude. On the other hand, low-latency WiFi technology can bring 5G level of service today if the network is properly set up to mitigate interference, given that distributed DCF per se does not impose inherent latency limitations in that it allows users to immediately access (after a short DIFS of 50 μs) the idle wireless channel in a decentralized manner.[7]

In this chapter, unlike Beyranvand et al. (2017) which studied only conventional H2H communication between MUs, we investigate the potential and limits of *coexistent teleoperation* in FiWi-enhanced LTE-A HetNets. Given the typical WiFi-only operation of state-of-the-art robots (Maier et al., 2016), HOs and TORs are assumed to communicate only via WLAN, as opposed to MUs who use dual-mode 4G/WiFi smartphones. Teleoperation is done either locally or nonlocally, depending on the proximity of the involved HO and TOR, as illustrated in Figure 2.11. In local teleoperation, the HO and corresponding TOR are associated with the same MAP and exchange their command and feedback samples through this MAP without traversing the fiber backhaul. Conversely, if HO and TOR are associated with different MAPs, nonlocal teleoperation is generally done by communicating via the backhaul EPON and central OLT. For simplicity, in this work, we focus on the generic network architecture of FiWi enhanced LTE-A HetNets, shown in Figure 2.11, without leveraging direct inter-ONU communication.

---

7 Aptilo Networks, "Why wait for 5G? Carrier Wi-Fi is here today," 22 December 2016. Online: www.wifinowevents.com

2.5 Delay Analysis

In this section, we develop our analytical framework to compute the average end-to-end delay and its distribution for local and nonlocal teleoperation with coexistent H2H traffic.

2.5.1 Assumptions

In our analysis, we make the following assumptions:

- *Single-hop WLAN*: MUs, HOs, and TORs are directly associated with an ONU-AP via a wireless single hop, whereby ONU-MPPs serve as ONU-APs (i.e. no MPs).

- *WiFi channel access*: Similar to Beyranvand et al. (2017), Kafaie et al. (2018), Medepalli and Tobagi (2006), Zhu et al. (2012), Liu et al. (2013), Han et al. (2006), Pham et al. (2005), and Aurzada et al. (2014), the WiFi channel access time governed by the IEEE 802.11 DCF is assumed to be exponentially distributed. This is justified by the DCF channel access mechanism, which includes carrier sensing, binary exponential back-off(s), and reattempts (if any) due to collisions and erroneous transmissions.

- *WiFi connectivity and WiFi offloading*: The WiFi connection and interconnection times of MUs are assumed to fit a truncated Pareto distribution, as validated via recent smartphone traces by Beyranvand et al. (2017). The probability $P_{\text{temporal}}^{\text{MU}}$ that an MU is temporarily connected to an ONU-AP is estimated as $\overline{T}_{\text{on}}/(\overline{T}_{\text{on}} + \overline{T}_{\text{off}})$, whereby $\overline{T}_{\text{on}}$ and $\overline{T}_{\text{off}}$ denote the average WiFi connection and interconnection time, respectively. In this chapter, we assume that $\overline{T}_{\text{on}} \gg \overline{T}_{\text{off}}$ based on the fact that the recent smartphone traces reported in by Beyranvand et al. (2017) indicate that the ratio $\overline{T}_{\text{on}}/(\overline{T}_{\text{on}} + \overline{T}_{\text{off}})$ has been constantly increasing. Hence, we assume that $P_{\text{temporal}}^{\text{MU}} \approx 1$ for MUs as well as HOs and TORs. Further, we assume that MUs offload their mobile traffic onto WiFi within the coverage area of an ONU-AP.

- *Traffic model*: MUs generate background Poisson traffic at mean packet rate λ_{BKGD} (in packets/s). Background traffic coming from ONUs with attached fixed (wired) subscribers is set to $\alpha_{\text{PON}} \cdot \lambda_{\text{BKGD}}$, where $\alpha_{\text{PON}} \geq 1$ is a traffic scaling factor for fixed subscribers that are directly connected to the backhaul EPON. Note that HOs and TORs generate traffic according to the different best fitting packet interarrival time distributions in Figure 2.8.

2.5.2 Local Teleoperation

For notational convenience, let us use the term "WiFi user" for all MUs, HOs, and TORs within the coverage area of an ONU-AP. We model each WiFi user as a GI/M/1 queue to account for the different packet interarrival time distributions under consideration. Let random variable D denote the delay experienced by any packet generated by a WiFi user, where D comprises queuing delay D_Q and service time D_S.

Suppose that packets arrive at rate λ at time instants $T_1, T_2, \ldots$, and assume that the interarrival times $T_{k+1} - T_k, k = 0, 1, \ldots$, are mutually independent, identically distributed random variables with distribution function $G(t) = P(T_{k+1} - T_k \leq t)$. Let N_k denote the number of packets in the system (i.e. queue and server) just prior to the arrival of packet k. By applying the theorem of total probability, we have

$$P(N_{k+1} = j) = \sum_{i=0}^{\infty} P(N_{k+1} = j \mid N_k = i)P(N_k = i), \quad j = 0, 1, 2, \ldots, \tag{2.3}$$
$$k = 0, 1, 2, \ldots$$

We define π_j as the probability that an arriving packet finds j packets in the system. A unique stationary distribution $\pi_j = \lim_{k \to \infty} P(N_k = j), j = 0, 1, 2, \ldots$, exists if and only if $\rho = \frac{\lambda}{\mu} < 1$, where μ denotes the service rate, which is equal to $1/\mathbb{E}[D_S]$. By taking limits on both sides of Eq. (2.3), we obtain

$$\pi_j = \sum_{i=0}^{\infty} p_{ij} \pi_i; \quad j = 0, 1, 2, \ldots \tag{2.4}$$

where $p_{ij} = P(N_{k+1} = j \mid N_k = i)$ represent the state transition probabilities and $\sum_{j=0}^{\infty} \pi_j = 1$. Clearly, we have $p_{ij} = 0, \forall j > i + 1$, because an arriving packet can find at most one more packet in the system than was found by the preceding packet. The remaining state transition probabilities can be computed by considering the following three cases:

Case 1: The server is busy between T_k and T_{k+1}, i.e. $i \geq 0$ and $1 \leq j \leq i + 1$. The probability that arriving packet $k + 1$ finds exactly j packets, given that the preceding packet k found i packets, is equal to the probability that exactly $i + 1 - j$ packets depart during interarrival time x. Thus, we have

$$P(N_{k+1} = j \mid N_k = i, T_{k+1} - T_k = x) = \frac{(\mu x)^{i+1-j}}{(i+1-j)!} e^{-\mu x}, \quad i \geq 0, 1 \leq j \leq i + 1 \tag{2.5}$$

Using interarrival time distribution function $G(x)$, we obtain

$$p_{ij} = \int_0^{\infty} \frac{(\mu x)^{i+1-j}}{(i+1-j)!} e^{-\mu x} dG(x), \quad i \geq 0, 1 \leq j \leq i + 1 \tag{2.6}$$

Case 2: The server becomes idle between two consecutive arrivals and arriving packet $k + 1$ and preceding packet k find the system empty, i.e. $i = j = 0$. This occurs if the service time of packet k is smaller than $T_{k+1} - T_k$. Hence, we have

$$p_{00} = \int_0^{\infty} (1 - e^{-\mu x}) dG(x) \tag{2.7}$$

Case 3: This case is like case 2 except that preceding packet k found $i \geq 1$ packets. The time y, $T_k < y < T_{k+1}$, until $(i + 1 - s)$th service completion leaving one packet in the system has an Erlang distribution with density function $f(y) = \frac{(\mu y)^{i-s}}{(i-s)!} e^{-\mu y} \mu$. For $y < x$, the probability of service completion during the remaining interarrival time interval of length $x - y$ equals $1 - e^{-\mu(x-y)}$. Thus, we have

$$p_{i0} = \int_0^\infty \int_0^x (1 - e^{-\mu(x-y)}) \frac{(\mu y)^{i-1}}{(i-1)!} e^{-\mu y} \mu \, dy \, dG(x), \quad i = 1, 2, \ldots \tag{2.8}$$

Lemma 2.1: *The stationary state probabilities π_j have a geometric distribution given by*

$$\pi_j = (1 - \omega)\omega^j \tag{2.9}$$

Proof: Let us consider the equilibrium probability state equations in Eq. (2.4) for $j \geq 1$. Substituting Eqs. (2.6) and (2.9) into Eq. (2.4) yields $\omega = \int_0^\infty e^{-(1-\omega)\mu x} \, dG(x)$, which can be rewritten as

$$\omega = \Phi(z)|_{z=(1-\omega)\mu} \tag{2.10}$$

where $\Phi(z)$ is the Laplace–Stieltjes transform of $G(x)$. The equation has a unique root in $(0, 1)$ if the queue is stable, i.e. $\rho < \lambda/\mu$. Substituting Eqs. (2.7)–(2.9) into Eq. (2.4) verifies Eq. (2.9) for $j = 0$.

Next, we compute the distribution of D_Q. Given N packets currently in the system, the probability $P(D_Q > t)$ is equal to $\sum_{i=1}^\infty \pi_i P(D_Q > t \mid N = i)$, which can be rewritten as

$$\sum_{j=0}^\infty (1 - \omega)\omega^{j+1} P(D_Q > t \mid N = j + 1)$$

The probability that a packet waits for longer than t in the queue given that $N = j + 1$ is equivalent to the probability that the number of departures during time interval t is smaller than or equal to j, which is given by

$$P(D_Q > t \mid N = j + 1) = \sum_{i=0}^j \frac{(\mu t)^i}{i!} e^{-\mu t} \tag{2.11}$$

Thus, we have

$$P(D_Q > t) = (1 - \omega)\omega \sum_{j=0}^\infty \omega^j \sum_{i=0}^j \frac{(\mu t)^i}{i!} e^{-\mu t} \tag{2.12}$$

which reduces to $P(D_Q > t) = \omega e^{-(1-\omega)\mu t}$. The cumulative distribution function (CDF) of D_Q is then given by

$$F_{D_Q}(t) = P(D_Q \leq t) = 1 - \omega e^{-(1-\omega)\mu t} \tag{2.13}$$

Next, let us consider D_S, whose CDF is given by

$$F_{D_S}(t) = P(D_S \leq t) = 1 - e^{-\mu t} \tag{2.14}$$

To compute the service rate μ in Eq. (2.14), we define the two-dimensional Markov process $(s(t), b(t))$ shown in Figure 2.12 under unsaturated non-Poisson traffic conditions and estimate the average service time $\mathbb{E}(D_S)$ in a WLAN using the IEEE 802.11 DCF for access control, whereby $b(t)$ and $s(t)$ denote the random backoff counter and size of contention window at time t, respectively. Let P_f and W_i denote the probability of a failed transmission attempt (i.e. collision or erroneous transmission) and contention window size at the back-off stage i, respectively. Note that the back-off stage i is incremented after each failed attempt up to the maximum value m, while the contention window is doubled at each stage, i.e. $W_i = 2^i W_0$.

A WiFi user is in idle state if (i) a successfully transmitted packet leaves the system without any waiting packet in the queue and (ii) no packet arrives during the current time slot given that the user was in idle state in the preceding time slot. We note that for non-Poisson arrival, these two events are not identical and should be calculated separately. Define π_j^* and $\hat{\pi}_j$ as the probability that a departing packet leaves j packets in the queue (i.e. from the viewpoint of the departing packet), and the fraction of time during which j packets are present in the queue (i.e. from the viewpoint of an outside observer), respectively. According to Figure 2.12, $1 - q_1$ is equal to the probability that a departing packet leaves the queue without any waiting packet, thus $1 - q_1 = \pi_0^*$. On the other hand, q_2 is the probability that at least one packet arrives during the current time slot given that the user was in idle state in the preceding time slot. This, however, does depend on the time interval during which the system has been in idle state so far. Nevertheless, for a slot duration being much smaller than the mean interarrival time, it is reasonable to estimate q_2 by $1 - \hat{\pi}_0 \approx \frac{\lambda}{\mu}$. Note that, according to Burke's theorem, $\pi_j^* = \pi_j$ holds for any arrival model, whereas $\hat{\pi}_j = \pi_j$ is valid only for Poisson arrival.

After finding the stationary distributions

$$b_{i,k} = \lim_{k \to \infty} P(s(t) = i, b(t) = k), \quad \forall k \in [0, W_i - 1], \quad i \in [0, m]$$

the probability τ that a WiFi user attempts to transmit in a given time slot is obtained as

$$\tau = \sum_{i=0}^{m} b_{i,0} = \cfrac{\cfrac{2(1 - 2P_f)q_2}{2(1 - q_1)(1 - P_f)(1 - 2P_f)}}{\cfrac{q_2[(W_0 + 1)(1 - 2P_f) + W_0 P_{eq}(1 - (2P_f)^m)]}{2(1 - q_1)(1 - P_f)(1 - 2P_f)} + 1} \tag{2.15}$$

The probability of a failed transmission attempt $P_{f,i}$ by WiFi user i is given by

$$1 - P_{f,i} = (1 - p_{e,i})(1 - p_{c,i}) \tag{2.16}$$

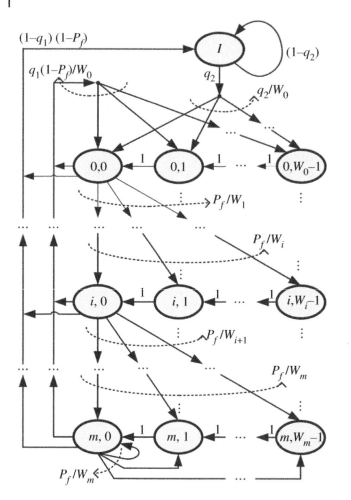

Figure 2.12 Two-dimensional Markov process. Source: Ebrahimzadeh and Maier (2019). © 2019 IEEE.

where $p_{e,i}$ and $p_{c,i}$ denote the probability of an erroneous transmission and the probability of a collision, respectively. Note that WiFi subscriber i does not experience a collision if the remaining users do not attempt to transmit, thus $1 - p_{c,i} = \prod_{v:v \neq i}(1 - \tau_v)$. Moreover, $p_{e,i}$ is estimated by $1 - (1 - p_b)^{\overline{L}_i}$, where $\overline{L}_i$ and p_b is the average length of a packet transmitted by WiFi user i and the bit error probability, respectively.

The probability of a collision-free packet transmission P_s given that there is at least one transmission attempt is given by $\frac{1}{P_{tr}} \left(\sum_i \tau_i \prod_{v,v \neq i}(1 - \tau_v) \right)$, whereby the probability P_{tr} that there is at least one transmission attempt is equal to

$1 - \prod_i (1 - \tau_i)$. The average slot duration E_s is then obtained as

$$E_s = (1 - P_{tr})\epsilon + P_{tr}(1 - P_s)T_c + P_{tr}P_sP_eT_e + P_{tr}P_s(1 - P_e)T_s \tag{2.17}$$

where T_c, T_e, and T_s are given in Aurzada et al. (2014). We then obtain

$$\mathbb{E}(D_S) = \frac{1}{\mu} = \sum_{k=0}^{\infty} p_e^k (1 - p_e)$$

$$\times \left[\sum_{j=0}^{\infty} p_c^j (1 - p_c) \cdot \left(\left(\sum_{b=0}^{k+j} \frac{2^{\min(b,m)}W_0 - 1}{2} E_s \right) + jT_c + kT_e + T_s \right) \right] \tag{2.18}$$

In order to obtain the steady-state values of q_1, q_2, P_f, τ, and μ, we numerically solve the system of nonlinear Eqs. (2.18), (2.16), (2.15), and (2.9).

The CDFs of D_Q (2.13) and D_S (2.14) are used to calculate the CDF of $D = D_Q + D_S$ at a WiFi user as follows

$$F_D(t) = P(D \leq t) = \int_0^t F_{D_S}(t - u)dF_{D_Q}(u) \tag{2.19}$$

The end-to-end delay of local teleoperation $D_{\text{LT}(i \rightarrow j)}^{\text{E2E}}$ between WiFi users i and j communicating via ONU-AP$_z$ is obtained as $D_{i \rightarrow \text{ONU-AP}_z} + D_{\text{ONU-AP}_z \rightarrow j}$, whose CDF is given by

$$F_{D_{\text{LT}(i \rightarrow j)}^{\text{E2E}}}(t) = P(D_{\text{LT}(i \rightarrow j)}^{\text{E2E}} \leq t) = \int_0^t F_{D_{i \rightarrow \text{ONU-AP}_z}}(t - \zeta)dF_{D_{\text{ONU-AP}_z \rightarrow j}}(\zeta) \tag{2.20}$$

where the CDFs $F_{D_{i \rightarrow \text{ONU-AP}_z}}(t)$ and $F_{D_{\text{ONU-AP}_z \rightarrow j}}(t)$ are calculated similar to Eq. (2.19).

2.5.3 Nonlocal Teleoperation

The average end-to-end delay of nonlocal teleoperation between WiFi user i and WiFi user j associated with ONU-AP$_m$ and ONU-AP$_n$, respectively, is given by

$$\overline{D}_{\text{NLT}(i \rightarrow j)}^{\text{E2E}} = \overline{D}_{i \rightarrow \text{ONU-AP}_m} + \overline{D}_{\text{PON}}^u + \overline{D}_{\text{PON}}^d + \overline{D}_{\text{ONU-AP}_n \rightarrow j} \tag{2.21}$$

where $\overline{D}_{i \rightarrow \text{ONU-AP}_m}$ and $\overline{D}_{\text{ONU-AP}_n \rightarrow j}$ denote the expected values of $D_{i \rightarrow \text{ONU-AP}_m}$ and $D_{\text{ONU-AP}_n \rightarrow j}$, respectively. Both expected values can be obtained from Eq. (2.19). Note that $\overline{D}_{\text{PON}}^u$ and $\overline{D}_{\text{PON}}^d$ denote the average delay of the backhaul EPON in the upstream and downstream direction, respectively, which are given by $\Phi(\rho^u, \overline{L}, \varsigma_L^2, c_{\text{PON}}) + \overline{L}/c_{\text{PON}} + 2\tau_{\text{PON}}\frac{2 - \rho^u}{1 - \rho^u} - B^u$ and $\Phi(\rho^d, \overline{L}, \varsigma_L^2, c_{\text{PON}}) + \overline{L}/c_{\text{PON}} + \tau_{\text{PON}} - B^d$, respectively; whereas ρ^u is the traffic intensity in upstream, ρ^d is the traffic intensity in downstream, τ_{PON} is the propagation delay between ONUs and OLT, c_{PON} is the EPON data rate, $\Phi(\cdot)$ denotes the well-known Pollaczek–Khinchine formula, and B^u and B^d are obtained as

$\Phi\left(\frac{\overline{L}}{\Lambda c_{PON}}\sum_{i=1}^{O}\sum_{q=1}^{O}\Gamma_{iq}^{PON}, \overline{L}, \varsigma_L^2, c_{PON}\right)$, where O is the number of ONUs and Γ_{iq}^{PON} is the traffic emanating from ONU$_i$ to ONU$_q$, and Λ denotes the number of wavelengths in the WDM PON (Beyranvand et al., 2017).

2.6 Edge Sample Forecast

Despite recent interest in exploiting machine learning for optical communications and networking, edge intelligence for enabling an immersive and transparent tele-operation experience for HOs has not been explored yet. In the following, we introduce machine learning at the edge of our considered communication network for realizing immersive and frictionless Tactile Internet experiences.

To realize edge intelligence, selected ONU-BSs/MPPs are equipped with AI-enhanced MEC servers. These servers rely on the computational capabilities of cloudlets collocated at the optical-wireless interface (see Figure 2.11) to forecast delayed haptic samples in the feedback path. Toward this end, we deploy a type of parameterized artificial neural network (ANN) known as *multilayer perceptron (MLP)*, which is capable of approximating any linear/nonlinear function to an arbitrary degree of accuracy (Hornik et al., 1989). Figure 2.13 illustrates the generic architecture of an MLP–ANN model. Note that an MLP with N_h hidden neurons represents a linear combination of N_h parameterized nonlinear functions called neurons. Furthermore, note that a neuron is a nonlinear function $\mathcal{G}(\cdot)$ of a linear combination of its input variables. In this work, the ANN is an MLP with L input variables and one output variable. More specifically, Ξ denotes the set of $L \cdot N_h + N_h + 1$ weights of the model, i.e. $\Xi = \{c_{i,j} : i = 1, \dots, N_h, j = 1, \dots, L\} \cup \{c_j' : j = 0, 1, \dots, N_h\}$, which are estimated during the training phase. The MLP yields the following output:

$$\Psi(\mathcal{A}, \Xi) = \sum_{j=1}^{N_h} c_j' \mathcal{G}\left(\sum_{i=1}^{L} c_{i,j} \mathcal{A}(i)\right) + c_0' \tag{2.22}$$

where $\mathcal{A} \in \mathbb{R}^L$ represents the input vector (see Figure 2.13). We note that the weights Ξ of the ANN are computed by the corresponding MEC server and are subsequently sent to the HO in close proximity.

Recall from Section 2.3.2 that deadband coding is less effective in the feedback path (see also Figure 2.7b). In this section, leveraging the notable amount of correlation between the haptic samples, as observed in our teleoperation traces in Section 2.3.3, we elaborate on our proposed ESF module as an interesting alternative to deadband coding in the feedback path, using the MLP described above instead. To do so, we present an ESF module based on the aforementioned MLP to compensate for delayed haptic feedback samples by means of

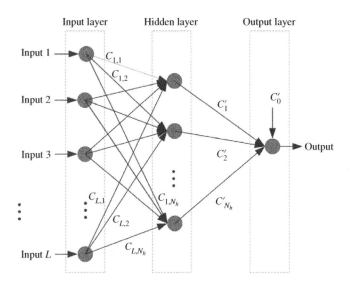

Figure 2.13 Generic architecture of a multilayer perceptron artificial neural network (MLP–ANN) model with L inputs and one output.

multiple-sample-ahead-of-time forecasting. As a result, the response time of the HO can be kept small, which in turn leads to a tighter togetherness with the remote TOR and an enhanced immersion. In a nutshell, our developed MLP based ESF module forecasts the force samples in the feedback path in real time. More specifically, instead of waiting for the force samples that are delayed by more than a given waiting deadline T_{thr}, the module locally generates and delivers the forecast feedback samples to the HO. Let us refer to the feedback signal to be forecast as the target signal $X(\cdot)$, i.e. the force feedback samples in our case. Our objective is to generate at any time instant t a forecast sample denoted by θ^* for time instant $t_0 = t - T_{\mathrm{thr}}$, whereby T_{thr} is the maximum period of time that the HO can wait until receiving the actual sample $\theta = X(t_0)$. More precisely, at any time t, if the sample for time instant t_0 is not received, a forecast sample is generated and immediately delivered to the HO. This procedure is repeated every 1 ms, which equals the typical intersample time of teleoperation systems. Note that the proposed MLP predicts θ from the past observations of the target signal. A technically more detailed description of our proposed ESF module is presented in the following.

Our objective is to generate at any time t a forecast sample θ^* for time instant $t_0 = t - T_{\mathrm{thr}}$, where T_{thr} is the waiting threshold until which the HO can wait to receive the actual sample $\theta = X(t_0)$. Let $\mathcal{S}, \mathcal{T} \in \mathbb{R}^K$ denote the last K samples $\{s_1, s_2, \ldots, s_K\}$ at time stamps $\{t_1, t_2, \ldots, t_K\}$. Note that $\mathcal{S}, \mathcal{T}$ are used to forecast

---

Algorithm 1 Edge Sample Forecast

Input: $\mathcal{T}, \mathcal{S}, t_0, \Xi$

Output: θ^*

1: $\delta = 1/F_s$

2: $\mathcal{T}^\delta, \mathcal{S}^\delta = \text{SAMPLE\_ALIGNER}(\mathcal{T}, \mathcal{S}, \delta)$

3: $\Delta \leftarrow \left\lceil \dfrac{t_0 - \mathcal{T}^\delta_{(L)}}{\delta} \right\rceil$

4: $\mathcal{A}_0 \leftarrow \left(s^\delta_1, \dots, s^\delta_L \right) \in \mathbb{R}^L$

5: **for** $i = 1$ to Δ **do**

6: $\quad t^*_i \leftarrow t^\delta_L + i \times \delta$

7: $\quad \theta_i = \Psi \left(\mathcal{A}_{i-1}, \Xi \right)$

8: $\quad \mathcal{A}_i = \left(\mathcal{A}_{i-1}(2), \mathcal{A}_{i-1}(3), \dots, \mathcal{A}_{i-1}(L), \theta_i \right)$

9: **end for**

10: $\theta^* \leftarrow \dfrac{\theta_\Delta - \theta_{\Delta-1}}{t^*_\Delta - t^*_{\Delta-1}} \left(t_0 - t^*_{\Delta-1} \right) + \theta_{\Delta-1}$

11: **return** θ^*

Source: Maier and Ebrahimzadeh (2019). © 2019 IEEE.

Algorithm 2 SAMPLE_ALIGNER()

Input: $\mathcal{T}, \mathcal{S}, \delta$

Output: $\mathcal{T}^\delta, \mathcal{S}^\delta$

1: $L \leftarrow \left\lceil \dfrac{t_K - t_1}{\delta} \right\rceil$

2: **for** $i = 1$ to L **do**

3: $\quad t^\delta_i \leftarrow t_1 + (i - 1)\delta$

4: **end for**

5: $s^\delta_1 \leftarrow s_1$

6: **for** $i = 2$ to L **do**

7: $\quad s^\delta_i \leftarrow \dfrac{s_j - s_{j-1}}{t_j - t_{j-1}} \left(t^\delta_i - t_{j-1} \right) + s_{j-1}, \forall j : t_{j-1} < t^\delta_i < t_j$

8: **end for**

9: **return** $\mathcal{T}^\delta, \mathcal{S}^\delta$

Source: Maier and Ebrahimzadeh (2019). © 2019 IEEE.

sample θ^* at any time instant $t_0 > t_K$. The feedback sample is forecast by Algorithm 1 with input $\mathcal{S}, \mathcal{T}, t_0, \Xi$ and output θ^*. We define δ as the intersample time step in our sample forecaster and set it to $1/F_s$, where F_s denotes the sampling frequency of 1 kHz (line 1 in Algorithm 1). To align the received samples in time, we call the SAMPLE_ALIGNER() procedure (Algorithm 2) with input $\mathcal{S} \in \mathbb{R}^K, \mathcal{T} \in \mathbb{R}^K$, and δ and output consisting of the aligned sample set $\mathcal{S}^\delta \in \mathbb{R}^L$ and time stamp set $\mathcal{T}^\delta \in \mathbb{R}^L$.

Next, we calculate the forecast horizon Δ at time t, which denotes the estimated number of samples during time interval $\mathcal{T}^\delta(L)$ between the last observed sample and target time t_0 (line 3 in Algorithm 1). Our objective is to forecast sample set $\Theta = \{\theta_1, \theta_2, \dots, \theta_\Delta\}$ for time stamp set $\{t_1^*, t_2^*, \dots, t_\Delta^*\}$ to finally estimate sample θ^* at time t_0. Specifically, sample $\theta_i \in \Theta$ is forecast by feeding our MLP with input vector $\mathcal{A}_{i-1} \in \mathbb{R}^L$, where $\mathcal{A}_0 = (s_1^\delta, \dots, s_L^\delta) \in \mathbb{R}^L$ and $\mathcal{A}_i = (\mathcal{A}_{i-1}(2), \mathcal{A}_{i-1}(3), \dots, \mathcal{A}_{i-1}(L), \theta_i)$, i.e. each sample is forecast based on the preceding L samples. To further improve the forecasting accuracy, we estimate θ^* by performing a two-point linear interpolation between $(t_{\Delta-1}^*, \theta_{\Delta-1})^T$ and $(t_\Delta^*, \theta_\Delta)^T$ (line 10 in Algorithm 1).

To create our training data set, we used the available 6-DoF teleoperation traces. We used MATLAB to build and train a one-hidden-layer ANN. Our training data set comprised 59 710 force feedback samples with the waiting deadline set to $T_{thr} = 1$ ms. We used the so-called "Levenberg–Marquardt" training method for adjusting the weights until a desired input/output relationship was obtained. Prior to simulations, we applied brute force for determining the optimal value of the number of neurons in the hidden layer, which led us to set it to 5. After training, we used a new data set comprising 1000 samples (different from the training data set) to evaluate the performance of our proposed sample forecaster in terms of mean squared error between the actual and forecast samples. It is worthwhile to mention that once the training was complete, the ANN was run on the HO side to provide the HO with forecast samples. Moreover, we note that the processing/running delay of the developed ANN on the order of microseconds was relatively small compared to the communication induced packet delays.

For completeness, we note that a one-hidden-layer MLP is also known as universal approximator. We decided to use a one-hidden-layer MLP since it is simple (i.e. easy to implement and train) yet achieves an accuracy that is good enough to approximate a wide variety of linear and/or nonlinear functions. Beside longer training times, note that increasing the number of hidden layers in our considered one-hidden-layer MLP may result in overfitting, which in turn may have a detrimental impact on its forecasting accuracy.

Note that in our considered FiWi-enhanced LTE-A HetNets architecture in Figure 2.11, all HOs and TORs are connected through a shared fiber backhaul, whose fiber reach does not exceed the typical 20 km of an IEEE 802.3ah EPON or up to 100 km in case of long-reach PONs. The limited fiber reach keeps the propagation delay below 0.1 and 0.5 ms, respectively. Thus, in a conventional EPON and in long-reach PONs the fiber propagation delay does not pose a challenge to meeting the 1 ms latency requirement of the Tactile Internet. However, an interesting question is how the 1-ms challenge of the Tactile Internet can be addressed for significantly larger geographical distances, e.g. connecting HOs in North America with TORs in Europe and/or Asia. This is where our proposed ESF module offers a potentially promising solution in that it decouples haptic feedback from the impact of extensive propagation delays, as typically encountered in wide area optical fiber networks. To see this, Figure 2.2

illustrates our ESF module for the general case of a communication network with arbitrary propagation delays. The ESF module may be inserted at the edge of the communication network in close proximity to the HO. Rather than waiting for delayed haptic feedback samples that exceed the waiting deadline of 1 ms, the ESF module generates forecast samples and delivers them to the HO. Hence, the HO is enabled to perceive the remote task environment in real time at a 1-ms granularity, resulting in a tighter togetherness, improved safety control, and increased reliability of the teleoperation systems. It should be noted, however, that a more rigorous experimental investigation would be needed to validate the viability of our proposed ESF module for real-word deployment scenarios with various wide area network propagation delays.

Clearly, the capability of our proposed ESF module to enable HOs to perceive the remote task environment in real time at a 1-ms granularity requires a sufficiently high forecasting accuracy of haptic feedback samples, as discussed in more detail later on.

2.7 Results

In this section, we present trace-driven simulation results along with numerical results derived from the analysis. Note that the obtained simulation results include confidence intervals at 95% confidence level. The following results were obtained by using the FiWi network parameter settings listed in Table 2.3. We assume that MUs, HOs, and TORs are directly connected to their associated MPPs, i.e. MPPs serve as conventional WLAN APs. By default, let us consider four ONU-APs, each with four associated MUs, whereby two MUs communicate with each other via their associated ONU-AP using an IEEE 802.11n WLAN (i.e. local H2H communications), while the two remaining MUs communicate with two uniformly randomly selected MUs associated with a different ONU-AP by using a backhaul IEEE 802.3ah 1Gb/s EPON with a typical fiber range of 20 km (i.e. nonlocal H2H communications). Furthermore, let us consider four conventional ONUs, serving fixed (wired) subscribers that are all involved in nonlocal H2H communications among each other. The MUs and fixed subscribers generate background traffic at a mean rate of λ_{BKGD} and $\alpha_{PON} \cdot \lambda_{BKGD}$, respectively. Note that $\alpha_{PON} \geq 1$ is a traffic scaling factor for fixed subscribers that are directly connected to the backhaul EPON. Figure 2.14 depicts the average end-to-end delay of MUs vs. mean background traffic rate λ_{BKGD} with different $\alpha_{PON} \in \{1, 50, 100\}$ for both local and nonlocal H2H communications in FiWi-enhanced LTE-A HetNets. The figure shows that an average end-to-end delay of $10^0 = 1$ ms can be achieved for nonlocal H2H communications for a wide range of background traffic loads.

Table 2.3 FiWi network parameters and default values.

Parameter	Value
Minimum contention window W_0	16
Maximum back-off stage H	6
Empty slot duration ϵ	9μs
DIFS	34μs
SIFS	16μs
PHY header	20μs
MAC header	36 bytes
RTS	20 bytes
CTS	14 bytes
ACK	14 bytes
Line rate r in wireless fronthaul	600 Mbps
Uplink and downlink data rate r_{PON} in PON	1 Gbps
l_{BKGD}	1500 bytes
p_b	10^{-6}
N_{DoF}	6
l_{PON}	20 km

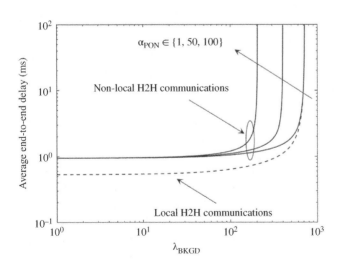

Figure 2.14 Average end-to-end delay of mobile users (MUs) vs. mean background traffic rate λ_{BKGD} (packets/s) for local and nonlocal human-to-human (H2H) communications with different $\alpha_{\mathrm{PON}} \in \{1, 50, 100\}$. Source: Maier and Ebrahimzadeh (2019). © 2019 IEEE.

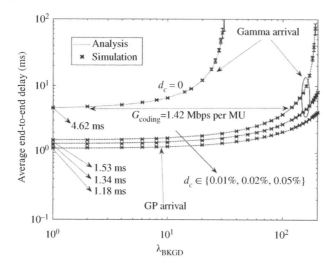

Figure 2.15 Average end-to-end delay of human operators (HOs) vs. mean background traffic rate λ_{BKGD} (packets/s) for local teleoperation with and without deadband coding in the command path for different $d_c \in \{0\%, 0.01\%, 0.02\%, 0.05\%\}$ ($\alpha_{PON} = 100$ fixed). Source: Maier and Ebrahimzadeh (2019). © 2019 IEEE.

Next, we include teleoperation and investigate the interplay between Tactile Internet traffic and the above H2H background traffic. Toward this end, we consider the above scenario and replace two MUs with a pair of HO and TOR in the coverage area of each ONU-AP for local teleoperation with and without deadband coding in the command path. Specifically, we consider our findings on 6-DoF teleoperation in Figure 2.8a and accordingly assume gamma and GP distributed haptic packet arrivals for $d_c \in \{0\%, 0.01\%, 0.02\%\}$ and $d_c = 0.05\%$, respectively. Figure 2.15 depicts the average end-to-end delay of HOs vs. mean background traffic rate λ_{BKGD} along with verifying trace-driven simulations based on our 6-DoF haptic traces and packetization procedure described in Section 2.3.1. We observe from Figure 2.15 that without deadband coding ($d_c = 0$) the minimum achievable average end-to-end delay experienced by HOs equals 4.62 ms, thus missing the Tactile Internet target of 1 ms. However, note that this target can be achieved with deadband coding for increasing d_c. For illustration, Figure 2.15 shows that we achieve a minimum average end-to-end delay of 1.18 ms for $d_c = 0.05\%$. In addition to decreasing the latency of HOs, note that deadband coding also has a beneficial impact on the admissible background traffic load of MUs due to the reduced haptic packet rates. To see this, let us define the coding gain G_{coding} as the difference between the maximum admissible throughput of MUs in teleoperation with and without deadband coding, while not violating a certain upper average end-to-end delay limit. For instance, for a given upper limit of 4.8 ms a coding

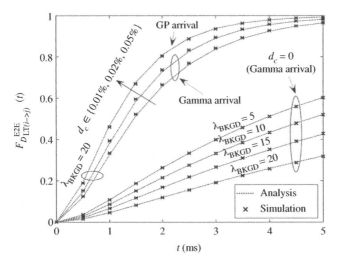

Figure 2.16 End-to-end delay cumulative distribution function (CDF) $F_{D_{LT(i\to j)}^{E2E}}(t)$ of local teleoperation. Source: Maier and Ebrahimzadeh (2019). © 2019 IEEE.

gain of $G_{coding} = 1.42$ Mbps per MU can be achieved in our teleoperation scenario by increasing d_c from 0% to 0.01%, as depicted in Figure 2.15. Note that overall the presented analytical results and verifying trace-driven simulation results (shown with 95% confidence interval) match very well.

Figure 2.16 provides useful insights into the upper end-to-end delay bounds by showing its CDF $F_{D_{LT(i\to j)}^{E2E}}(t)$ for the scenario of Figure 2.15. Notably, we observe that for $d_c = 0.05\%$ and a high background traffic rate of $\lambda_{BKGD} = 20$ packets/s (top curve), the end-to-end delay stays below 2 ms with a probability as high as 0.8.

To provide insights into the impact of different NG-PON backhaul infrastructures in the case of nonlocal teleoperation, Figure 2.17 depicts the average end-to-end delay performance of HOs vs. backhaul traffic scale factor α_{PON} of fixed subscribers with the mean background traffic rate set to $\lambda_{BKGD} = 20$ packets/s. For comparison, we consider a conventional 1 Gbps EPON, a high-speed 10 Gbps EPON, and a WDM PON with $\Lambda = 2$ wavelength channels, each operating at 1 Gb/s. Note that for all three considered NG-PONs we include a conventional fiber reach of $l_{PON} = 20$ km as well as its respective long-reach counterpart with an extended fiber reach of $l_{PON} = 100$ km. We observe from Figure 2.17 that the use of deadband coding ($d_c = 0.05\%$) is instrumental in lowering the average end-to-end delay below 10 ms for all NG-PON backhaul infrastructures under consideration. The figure also confirms previous findings (see Section 2.1) that 10G PON and WDM technologies represent cost-effective solutions to support 5G

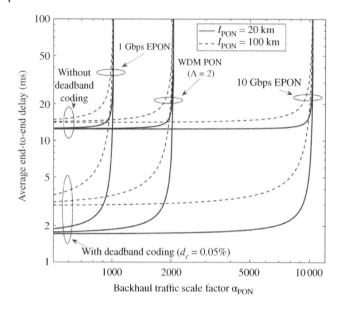

Figure 2.17 Average end-to-end delay of human operators (HOs) vs. backhaul traffic scale factor α_{PON} of fixed subscribers ($\lambda_{BKGD} = 20$ packets/s fixed) for nonlocal teleoperation across different NG-PON backhaul infrastructures. Source: Maier and Ebrahimzadeh (2019). © 2019 IEEE.

low-latency applications over a wide range of backhaul traffic loads by sharing a common optical transport platform among fixed subscribers, MUs, and HOs.

We have seen in the results above that deadband coding is effective in decreasing the average end-to-end delay by reducing the haptic packet rate. Nevertheless, some haptic packets may still experience an instantaneous delay that exceeds the desired waiting deadline on the order of 1 ms until their reception due to varying traffic conditions and MAC layer queuing times. To ensure that the HO receives expected haptic packets before the deadline, our proposed MLP based ESF module may be used as a complementary technique to deadband coding in the feedback path. Figure 2.18 compares the forecasting accuracy of our proposed MLP based ESF scheme with a naive ESF scheme, where the forecast sample is simply set to the last received sample. In our simulation, we used our 6-DoF teleoperation traces to train a one-hidden-layer MLP by using 59 710 force feedback samples with the waiting deadline set to $T_{thr} = 1$ ms. Figure 2.18 clearly shows the superior forecasting accuracy of our proposed MLP based ESF scheme in terms of mean squared error over a wide range of λ_{BKGD} for both local and nonlocal teleoperation scenarios, whereby a low mean squared error is achievable in the former scenario. Specifically, for nonlocal teleoperation, our MLP-based ESF scheme

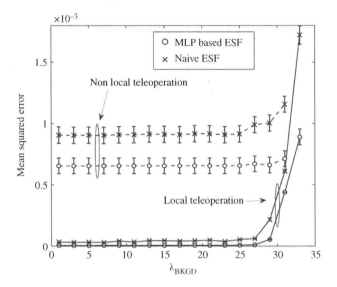

Figure 2.18 Comparison of forecasting accuracy between proposed multilayer perceptron (MLP) based and naive edge sample forecast (ESF) schemes for local and nonlocal teleoperation without deadband coding in the feedback path ($d_f = 0$). Source: Maier and Ebrahimzadeh (2019). © 2019 IEEE.

decreases the mean squared error from roughly 0.9 to 0.65 $\times 10^{-3}$, translating into an improvement of 27.8%. For local teleoperation, it is able to keep the mean squared error close to zero between 0.006 and 0.007 $\times 10^{-3}$ at a low to medium background traffic load λ_{BKGD}. Note that the observed performance improvement is due to the relatively high autocorrelation in the haptic feedback samples that allows our proposed MLP based ESF module to achieve a more accurate forecast compared to that of the naive ESF scheme.

2.8 Conclusions

We have seen that there is a significant overlap among 5G, IoT and the Tactile Internet in that they share various important design goals, including very low latency, ultrahigh reliability, and integration of data-centric technologies. This chapter described how FiWi-enhanced LTE-A HetNets leveraging low-cost data-centric EPON and WiFi technologies for fiber backhaul sharing and WiFi offloading may help realize not only the aforementioned shared design goals but also the key attributes of end-to-end co-DBA of both PON and wireless network resources, decentralization, and edge intelligence in support of 5G low-latency applications over a common optical transport platform.

Our focus was on the emerging Tactile Internet as one of the most interesting 5G low-latency applications for creating novel immersive experiences. Recall from Chapter 1 that the emerging Tactile Internet will remain a prominent application enabled by future 6G mobile networks. We reviewed the HART-centric design principles that add a new dimension to the H2M interaction via the Internet and set the Tactile Internet aside from the more machine-centric IoT. Exploiting the human perception of haptics to reduce the haptic packet rate by means of deadband coding, we derived haptic traffic models from teleoperation experiments. Our haptic trace analysis showed that assuming Tactile Internet traffic to be Pareto distributed was not valid for the analyzed traffic, while assuming it to be Poisson traffic was valid only in a special case. In general, we observed that command and feedback paths of teleoperation systems can be jointly modeled by GP, gamma, or deterministic packet interarrival time distributions, depending on the given value of the respective deadband parameters.

We elaborated on the importance of the decentralized nature of WLAN's access protocol DCF to realize low-latency FiWi enhanced LTE-A HetNets. Furthermore, by exploiting their inherent distributed processing and storage capabilities, we investigated the potential of enabling immersive teleoperation experiences for HOs by introducing machine learning at the optical-wireless interface of FiWi-enhanced LTE-A HetNets. Our proposed MLP based ESF module compensates for delayed haptic feedback samples by means of multiple-sample-ahead-of-time forecasting for a tighter togetherness, improved safety control, and increased reliability.

3

Context- and Self-Awareness for Human-Agent-Robot Task Coordination

3.1   Introduction

Today's telecommunication networks enable us to connect devices and people for an unprecedented exchange of audiovisual and data content. With the advent of commercially available haptic/tactile sensory and display devices, conventional triple-play (i.e. audio, video, and data) content communication now extends to encompass the real-time exchange of haptic information (i.e. touch and actuation) for the remote control of physical and/or virtual objects through the Internet. This paves the way toward realizing the so-called *Tactile Internet* (Maier et al., 2016), whereby human–machine interactions will convert today's content-delivery networks into skillset/labor-delivery networks (Aijaz et al., 2017). The Tactile Internet holds great promise to have a profound socioeconomic impact on a broad array of applications in our everyday life, ranging from industry automation and transport systems to healthcare, telesurgery, and education (Maier and Ebrahimzadeh, 2019).

Beside the design of low-latency/jitter and highly reliable networking infrastructures, a key challenge little discussed in the existent Tactile Internet literature is how we can make sure that the potential of the Tactile Internet be unleashed for a race with (rather than against) machines. Recall from Chapter 2 that the overarching goal of the Tactile Internet should be the production of new goods and services by means of empowering rather than automating machines that complement humans rather than substitute for them. We note that any technological advance can be labor-saving or capital-saving. In either case, regardless of the speed with which robots approach or even exceed human skill sets, the key to the effect of the new technologies on human society is who owns the technologies. We would lose our jobs if other persons owned our replacement technologies. By contrast, if users

Toward 6G: A New Era of Convergence, First Edition. Amin Ebrahimzadeh and Martin Maier.
© 2021 The Institute of Electrical and Electronics Engineers, Inc.
Published 2021 by John Wiley & Sons, Inc.

owned them, humans would have their current earnings and their time freed from labor to explore other productive activities (Freeman, 2016).

In this chapter, we leverage on our recently proposed concept of FiWi enhanced long-term evolution-advanced (LTE-A) HetNets, which were shown to achieve the 5G and Tactile Internet key requirements of very low latency on the order of 1–10 ms and ultra-high reliability by unifying coverage-centric 4G mobile networks and capacity-centric FiWi broadband access networks based on low-cost, data-centric Ethernet next-generation passive optical network (NG-PON) and Gigabit-class wireless local area network (WLAN) technologies (Beyranvand et al., 2017). While necessary, though, the design of reliable low-latency converged communication network infrastructures is not sufficient to realize the full potential of the Tactile Internet.

Depending on the context-awareness[1] of future Tactile Internet applications, tasks may be classified into three different categories: (i) location-dependent physical-only tasks (e.g. lifting an object), (ii) location-independent digital-only tasks (e.g. object classification from a captured image, which might be offloaded for computation at a remote cloud or nearby cloudlet), or (iii) location-dependent physical/digital tasks that include both types of tasks (e.g. assemblage followed by a unit test). As users will need to request robot assistance from time to time, mapping these requests to the robots stands as an optimization problem, whose objective is to minimize not only the task completion time but also the operational expenditure (OPEX) and robot energy consumption. The difficulty of solving such a problem lies in the following reasons. First, it is clear that we are dealing with different conflicting objectives, which makes it challenging to obtain a satisfactory result, especially for large-sized problems. Second, in real-world scenarios, there is no a priori knowledge of the task arrival times, making it almost impossible to obtain the optimal solution. Third, and more importantly, to minimize the energy consumption of mobile robots (MRs), the task coordinator requires global knowledge of all the system parameters, including in particular the local parameters of the MRs, which may not be willing to share such private information.

In this chapter, we use context-awareness to develop a HART-centric task coordination algorithm that minimizes the completion time of physical/digital tasks as well as OPEX by spreading ownership of robots across mobile users. In addition, we capitalize on self-awareness[2] to improve the performance of a given robot

---

1 Context refers to the information that can be used to characterize situation of a relevant entity. Accordingly, context-awareness is the ability to adapt behavior according to changes in surroundings.
2 Self-awareness is the ability of networks to observe their own internal status, objectives, and preferences and to modify their internal behavior so as to adaptively achieve certain goals, e.g. compensating for failing or malfunctioning components.

by identifying its capabilities as well as the objective requirements by means of optimal motion planning to minimize its energy consumption as well as traverse time. Our proposed self- and context-aware HART-centric allocation scheme for both physical and digital tasks is used to coordinate the automation and augmentation of mutually beneficial human–machine coactivities across a FiWi-based Tactile Internet infrastructure. In particular, the contributions of this chapter are as follows:

- We formulate a multiobjective optimization problem to minimize the task completion time, energy consumption, and OPEX for multirobot task allocation in the Tactile Internet over FiWi-enhanced networks.
- We develop a context-aware HART-centric task coordination algorithm that minimizes the completion time of physical/digital tasks, while paying particular attention to reducing OPEX by spreading ownership of robots across mobile users.
- We propose a self-aware optimal motion planning algorithm, which runs locally at the MRs, with the objective to find the best trade-off between traverse time and energy consumption by leveraging on *local self-awareness* of the MRs to identify their respective limitations and capabilities as well as objective requirements for accomplishing the allocated tasks.
- We provide an analytical framework to calculate the average packet transmission delay and human–robot connection reliability, two key attributes of the Tactile Internet.

The remainder of the chapter is structured as follows. Section 3.2 describes our considered FiWi-based Tactile Internet infrastructures for HART-centric task coordination in greater detail, followed by motion and energy consumption models for MRs. In Section 3.3, we develop our multiobjective optimization problem considering characteristics and key parameters of MRs and tasks, which is subsequently solved by our proposed HART-centric context-aware multi-robot task coordination algorithm. In Section 3.4, we present our self-aware optimal motion planning algorithm. Our delay and reliability analysis is presented in Section 3.5. In Section 3.6, we report on our obtained results and findings. Section 3.7 concludes the chapter.

3.2 System Model

3.2.1 Network Architecture

Figure 3.1 illustrates the generic network architecture of our considered FiWi-enhanced LTE-A HetNets. The optical backhaul consists of a time division

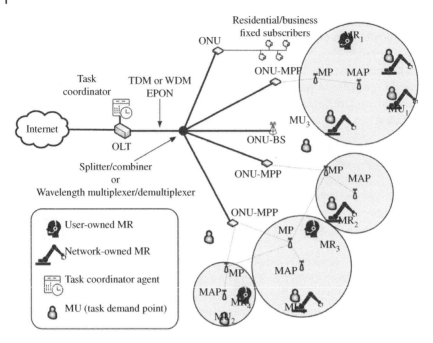

Figure 3.1 Generic architecture of fiber-wireless (FiWi) based Tactile Internet network infrastructure for multirobot task coordination. Source: Ebrahimzadeh et al. 2019. © 2019 IEEE.

multiplexing (TDM)/wavelength division multiplexing (WDM) IEEE 802.3ah/av 1/10 Gb/s Ethernet passive optical network (EPON) with a typical fiber length of 20 km between the central optical line terminal (OLT) and remote optical network units (ONUs). The EPON may comprise multiple stages, each stage separated by a wavelength-broadcasting splitter/combiner or a wavelength multiplexer/demultiplexer. There are three different subsets of ONUs. An ONU may either serve fixed (wired) subscribers. Alternatively, it may connect to a cellular network base station (BS) or an IEEE 802.11n/ac/s WLAN mesh portal point (MPP), giving rise to a collocated ONU-BS or ONU-MPP, respectively. Depending on her trajectory, an mobile user (MU) may communicate through the cellular network and/or WLAN mesh front-end, which consists of ONU-MPPs, intermediate mesh points (MPs), and mesh access points (MAPs).

Note that tasks arrive at random time instants at the MUs, which act as the demand points. The MUs then send their demands upstream to the task coordinator agent, which is colocated with the OLT (see also Figure 3.1), via the wireless front-end and EPON backhaul until they reach the OLT. The task

Figure 3.2 Trapezoidal velocity profile of mobile robots (MRs). Source: Ebrahimzadeh et al. 2019. © 2019 IEEE.

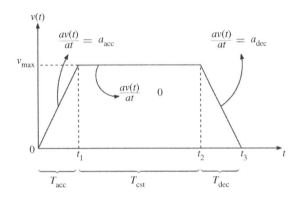

coordinator agent is responsible for allocating the incoming tasks to MRs,[3] which may be owned by either the users or the network operator. After receiving the task allocation from the OLT, the selected MR moves toward the demand point and collaboratively executes the physical and/or digital tasks. After successfully executing the tasks, the MR transmits the result/output of the physical and/or digital task to the task demand point (i.e. MU).

3.2.2 Energy and Motion Models of Mobile Robots

For an MR with forward translational velocity powered by a direct current (DC) motor, we use the detailed model presented in Tokekar et al. (2014). Specifically, let $v(t)$ and $a(t)$ denote the velocity and acceleration profile of the MR, respectively. The energy consumption of the brushed DC motor deployed at the MR is given by

$$E = \int_0^{t_3} e(t)i(t)dt = \int_0^{t_3} [c_1 a^2(t) + c_2 v^2(t) + c_3 v(t) + c_4 + c_5 a(t) + c_6 v(t)a(t)]dt$$

(3.1)

where $i(t)$ and $e(t)$ denote the instantaneous current and voltage of the DC motor, respectively, while the constants $\{c_i\}_{i=1}^6$, given by Tokekar et al. (2014), are combinations of the motor parameters and depend on the design of the motor and surface on which the robot traverses.

Next, let us consider the trapezoidal velocity profile of the MR, shown in Figure 3.2. The profile indicates that along a given path the MR accelerates from rest during T_{acc}, traverses with constant velocity v_{max} during T_{cst}, and then decelerates during T_{dec} until it returns to rest. Based on the considered velocity

---

3 The reader should that even though the acronym MR is widely referred to "mixed reality," we have frequently used MR standing for "mobile robot" only within this chapter.

profile, the distance Δd traversed by the MR is given by

$$\Delta d = \int_0^{T_{\text{trav}}} v(t)dt = (T_{\text{acc}} + 2T_{\text{cst}} + T_{\text{dec}})\frac{v_{\max}}{2} \tag{3.2}$$

which yields

$$T_{\text{acc}} + T_{\text{cst}} = \frac{\Delta d}{v_{\max}} \tag{3.3}$$

by assuming $a_{\text{acc}} = -a_{\text{dec}}$. Having

$$T_{\text{cst}} = \omega_d \frac{\Delta d}{v_{\max}} \tag{3.4a}$$

$$T_{\text{acc}} = T_{\text{dec}} = (1 - \omega_d)\frac{\Delta d}{v_{\max}} \tag{3.4b}$$

the traverse time T_{trav} is then equal to

$$\begin{aligned}
T_{\text{trav}} &= (1 - \omega_d)\frac{\Delta d}{v_{\max}} + \omega_d \frac{\Delta d}{v_{\max}} + (1 - \omega_d)\frac{\Delta d}{v_{\max}} \\
&= (2 - \omega_d)\frac{\Delta d}{v_{\max}}
\end{aligned} \tag{3.5}$$

with $\omega_d \in [0, 1)$. Clearly, T_{trav} is a monotonically decreasing function of ω_d with the upper and lower bounds given by

$$T_{\text{trav}}^U = \lim_{\omega_d \to 0} T_{\text{trav}} = 2\frac{\Delta d}{v_{\max}} \tag{3.6a}$$

$$T_{\text{trav}}^L = \lim_{\omega_d \to 1} T_{\text{trav}} = \frac{\Delta d}{v_{\max}} \tag{3.6b}$$

Lemma 3.1 For the velocity profile shown in Figure 3.2, the energy consumption E_{trav} of the MR to traverse a given distance Δd is given by

$$E_{\text{trav}} = E(\omega_d) = \frac{2c_1 v_{\max}^3}{(1 - \omega_d)\Delta d} + \frac{c_2}{3}(\omega_d + 2)v_{\max}\Delta d + c_3\Delta d + (2 - \omega_d)\frac{c_4\Delta d}{v_{\max}} \tag{3.7}$$

Proof: See Appendix A.1. □

Note that $E(\omega_d)$ is a convex function of ω_d, as $\frac{\partial^2 E(\omega_d)}{\partial \omega_d^2} = \frac{4c_1 v_{\max}^3}{\Delta d(1-\omega_d)^3} > 0$ for $\omega_d \in [0, 1)$.

Lemma 3.2 $E(\omega_d)$ has a local minimum in interval $(0, 1)$ if and only if

$$v_{\max} < \sqrt{\frac{-c_2 + \sqrt{c_2^2 - 4\left(\frac{6c_1}{\Delta d^2}\right)(-3c_4)}}{2\left(\frac{6c_1}{\Delta d^2}\right)}} \tag{3.8}$$

otherwise, $E(\omega_d)$ is a monotonically increasing function of ω_d with a minimum at $\omega_d = 0$.

Proof: See Appendix A.2. □

3.3 Context-Aware Multirobot Task Coordination

In this section, we study the problem of task allocation to MRs in multirobot FiWi-based infrastructures in greater detail. We note that the automation of various physical and digital tasks with context-aware requirements is doable by state-of-the-art agents and robots. We start by presenting an illustrative use case as a simplified example of our optimization problem of interest. Next, we develop the multiobjective formulation of our problem. We then develop a context-aware allocation algorithm of physical/digital tasks for the HART-centric multirobot task coordination based on the shared use of user- and network-owned robots.

3.3.1 Illustrative Case Study

For illustration, we present a case study to better understand the impact of different coordination strategies on the delay/cost/energy performance from the viewpoint of both users and network operator. Note that multirobot systems have attracted attention in a wide variety of applications such as exploration (Elizondo Leal et al., 2016), tracking (Chang et al., 2016), foraging (Lee et al., 2014), transportation (Barrientos et al., 2016), and manufacturing (Tereshchuk et al., 2019). Let us consider two user- and three network-owned MRs (i.e. a 40% user ownership), as shown in Figure 3.3, where a task has arrived at the demand point to be allocated to one of the MRs. The next available time of each MR is also shown. Assume that receiving service from a user- and network-owned MR is subject

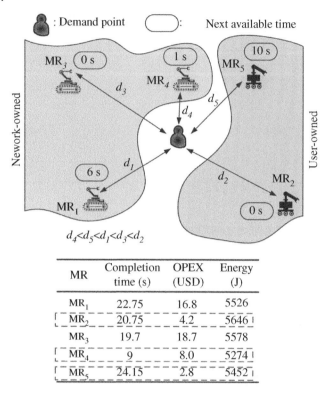

Figure 3.3 An illustrative case study demonstrating the trade-off between delay, OPEX, and energy performance of the multirobot task allocation problem at a given task arrival time instant. Source: Ebrahimzadeh et al. 2019. © 2019 IEEE.

to an incurred OPEX of 0.2 and 1 USD per second, respectively. The outcome of the allocation of the given task to each of the MRs is shown in Figure 3.3, which demonstrates the trade-off between task completion time, OPEX, and energy consumption. The results indicate that allocating the task to MR_2, MR_4, or MR_5 is a Pareto optimal solution with respect to the three objectives of task completion time, OPEX, and energy consumption, i.e. one cannot find any other solution whose performance in terms of all the three objectives is better than MR_2, MR_4, or MR_5. Note that any allocation decision made for a given task updates the next available time of the allocated MR, thus having a direct impact on the performance results for upcoming tasks, whose arrival time instants are not known in advance.

3.3.2 Problem Formulation

We assume that HART members are self-aware about their respective goals, application needs, capabilities, and constraints to be elaborated on in Section 3.4.

Further, through communication, they establish a collective context-awareness with the objective of minimizing the completion time of tasks by MRs, which may be either user-owned or network-owned. Let the ownership spreading factor γ_O denote the percentage of robots that are jointly owned by MUs, whereas the remaining robots are owned by the network operator. More specifically, our multirobot task coordination algorithm aims to minimize the task completion time along with the energy consumption and OPEX of physical/digital task execution by MRs. In the following, after introducing the decision variables and parameters, we develop a multiobjective formulation of the dynamic task allocation problem.

Given:

- J_i: Task i, $(i = 1, 2, \ldots)$.
- t_i^a: Arrival time of task demand i.
- W_i^p: Physical workload (in Joules) generated by J_i.
- W_i^d: Digital workload (in required central processing unit [CPU] cycles) generated by J_i.
- l_i^{task}: Demand location of task J_i.
- $\mathcal{R}_N$: Set of network-owned MRs.
- $\mathcal{R}_U$: Set of user-owned MRs.
- $\mathcal{R}_U^A$: Set of available user-owned MRs.
- $\mathcal{R}_U^B$: Set of busy user-owned MRs.
- $\mathcal{R}_N^A$: Set of available user-owned MRs.
- $\mathcal{R}_N^B$: Set of busy network-owned MRs.
- $\mathcal{R}$: Set of all user- or network-owned MRs.
- N: Total number of MRs.
- l_j^r: Location of MR$_j$.
- t_j^{av}: Next available time of MR$_j$.
- v_{max}^j: Maximum speed of robot MR$_j$.
- $a_{\text{acc}}^{\text{max},j}$: Maximum acceleration of robot MR$_j$.
- C_j^p: Physical task processing capacity (in Watts) of MR$_j$.
- C_j^d: Digital task processing capacity (in CPU cycles per time unit) of MR$_j$.
- D: Maximum scheduling deadline.
- $d(l_j^r, l_i^{\text{task}})$: Euclidean distance between the demand location of task J_i and MR$_j$.

Parameters:

- φ_U: Operational cost per time unit of user-owned MRs.
- φ_N: Operational cost per time unit of network-owned MRs.
- ϵ_d: Energy (in Joules) per CPU cycle.

Decision variables:

- X_i^j: A binary variable set to 1 if task J_i is assigned to MR_j.

Objectives:

- $T(\mathbf{X})$: Task completion time.
- $C(\mathbf{X})$: Operational expenditures (OPEX).
- $E(\mathbf{X})$: Energy consumption.

Multiobjective formulation:

$$
\begin{aligned}
&\underset{\mathbf{X}}{\text{minimize}} && T(\mathbf{X}), C(\mathbf{X}), E(\mathbf{X}), && \forall i = 1, 2, 3, \ldots \\
&\text{subject to} && \sum_{j \in R_U} \max\{t_j^{av} - t_i^a, 0\}X_i^j < D \\
& && \sum_{j=1}^{N} X_i^j = 1 \\
& && X_i^j \in \{0, 1\}, \quad \forall j = 1, 2, \ldots, N
\end{aligned}
\tag{P1}
$$

where $T(\mathbf{X})$, $C(\mathbf{X})$, and $E(\mathbf{X})$ are obtained as follows. The total task completion time comprises the following delay components: (i) transmission delay T_{trs}^{dmd} of allocation demand from a given MU to the OLT, (ii) scheduling delay $T_{sch}^{i,j}$, which is the elapsed time between arrival time t_j^a of task J_i until MR_j becomes available, (iii) transmission delay T_{trs}^{alc} of allocation from the OLT to the allocated MR, (iv) traverse time $T_{trav}^{i,j}$, which is the amount of time that takes MR_j to traverse to the demand location of task J_i, (v) execution time $T_{exc}^{i,j}$, which is the amount of time that takes MR_j to execute physical/digital task J_i, and (vi) transmission delay T_{trs}^{o} to transmit the output/result of digital/physical task from the MR to the MU. $T(\mathbf{X})$ is then given by

$$
T(\mathbf{X}) = \sum_{j=1}^{N} X_i^j (T_{trs}^{dmd} + T_{sch}^{i,j} + T_{trs}^{alc} + T_{trav}^{i,j} + T_{exc}^{i,j} + T_{trs}^{o}), \quad \forall i = 1, 2, \ldots
\tag{3.9}
$$

where the scheduling delay $T_{sch}^{i,j}$ is obtained as follows:

$$
T_{sch}^{i,j} = \max\{t_j^{av} - t_i^a, 0\}, \quad \forall i = 1, 2, \ldots, \forall j = 1, 2, \ldots, N
\tag{3.10}
$$

We note that after rearranging and considering $\sum_{N}^{j=1} X_i^j = 1, \forall i = 1, 2, \ldots$, Eq. (3.9) reduces to

$$
T(\mathbf{X}) = T_{trs}^{dmd} + T_{trs}^{alc} + T_{trs}^{o} + \sum_{j=1}^{N} X_i^j (T_{sch}^{j} + T_{trav}^{j} + T_{exc}^{j}), \quad \forall i = 1, 2, \ldots
\tag{3.11}
$$

Before estimating the task execution time, let incoming task J_i consist of both physical and digital workloads denoted by W_i^p (in Joules) and W_i^d (in required CPU cycles), respectively. We then estimate the task execution time $T_{\text{exc}}^{i,j}$ by

$$T_{\text{exc}}^{i,j} = \underbrace{\frac{W_i^p}{C_j^p}}_{\text{physical sub-task}} + \underbrace{\frac{W_i^d}{C_j^d}}_{\text{digital sub-task}}, \quad \forall i = 1, 2, \ldots \tag{3.12}$$

The traverse time $T_{\text{trav}}^{i,j}$ is given in Eq. (3.5), whereas the other delay components that are related to packet transmission delay will be computed shortly in Section 3.5.1.

Next, we estimate the OPEX of task execution by user- and/or network-owned MRs. To do so, we assume a flat-rate pricing policy that charges MUs from the time instant when the MR becomes available and is allocated to the task until it successfully accomplishes task execution. Let φ_U and φ_N denote the operating cost per time unit for user- and network-owned MRs, respectively, whereby $\frac{\varphi_U}{\varphi_N} \leq 1$. We then estimate the OPEX, $C(\mathbf{X})$, of task execution as follows:

$$C(\mathbf{X}) = \sum_{j \in S_U} \varphi_U X_j^i (T_{\text{trav}}^{i,j} + T_{\text{exc}}^{i,j}) + \sum_{j \in S_N} \varphi_N X_j^i (T_{\text{trav}}^{i,j} + T_{\text{exc}}^{i,j}), \quad \forall i = 1, 2, \ldots \tag{3.13}$$

Next, let us calculate the total energy consumption. We note that the energy consumed to transmit the task demand/allocation/output/result is negligible compared to the energy consumption of an MR to traverse and execute the task. Therefore, we consider only the energy consumption of MRs to traverse to the demand location and execute the physical/digital task. Accordingly, we model the total energy consumption, $E(\mathbf{X})$, as follows:

$$E(\mathbf{X}) = \sum_{j=1}^{N} X_j^i (E_{\text{trav}}^{i,j} + E_{\text{exc}}^i) \tag{3.14}$$

where $E_{\text{trav}}^{i,j}$ is given in Eq. (A.1) and execution energy E_{exc}^i of task J_i (which is independent of the MR selection) is given by

$$E_{\text{exc}}^i = \underbrace{W_i^p}_{\text{physical sub-task}} + \underbrace{\epsilon_d W_i^d}_{\text{digital sub-task}}, \quad \forall i = 1, 2, \ldots \tag{3.15}$$

where ϵ_d denotes the energy (in Joules) per CPU cycle. Note that $E_{\text{trav}}^{i,j}$ depends on the MR selection, while E_{exc}^i does not. Thus, Eq. (3.14) reduces to

$$E(\mathbf{X}) = E_{\text{exc}}^i + \sum_{j=1}^{N} X_j^i E_{\text{trav}}^{i,j}, \quad \forall i = 1, 2, \ldots \tag{3.16}$$

3.3.3 The Proposed Algorithm

Clearly, $T(\mathbf{X})$, $C(\mathbf{X})$, and $E(\mathbf{X})$ may be conflicting objectives, as minimizing $T(\mathbf{X})$ and $E(\mathbf{X})$ may not necessarily minimize $C(\mathbf{X})$ (see Figure 3.3). The reason for this is that for some tasks selecting network-owned MRs can significantly reduce the task completion time, resulting in increased OPEX due to higher pricing of network-owned MRs compared to that of user-owned ones. We also note that the energy consumption of an MR is a function of its local parameters, e.g. motor and motion parameters, among others, which may preferably not be shared by the MRs, as they are considered private information. Furthermore, the task coordinator has to make decisions without a priori knowledge of the arrival time instants of upcoming tasks, thus making it impossible to exploit conventional optimization methods to obtain the optimal solution of the problem of interest. Therefore, in order to make a suitable trade-off between the three objectives and achieve a satisfactory solution, we prioritize the objectives of the problem in descending order of $T(\mathbf{X})$, $C(\mathbf{X})$, and $E(\mathbf{X})$. More specifically, we decouple the problem into two subproblems namely multirobot task coordination and motion planning, where the former aims to minimize $T(\mathbf{X})$ and $C(\mathbf{X})$, whereas the latter minimizes $E(\mathbf{X})$ (to be discussed in Section 3.4).

As shown in Algorithm 3, our proposed context-aware dynamic multirobot task coordination (CADMRTC) algorithm assigns the given task to the nearest available user-owned MR, if there is any (see line 2 of Algorithm 3). Otherwise, it tries to find the earliest available user-owned MR up to a given maximum scheduling deadline $D \geq 0$ seconds before falling back onto network-owned MRs (see lines 6–15 of Algorithm 3). In this case, the task is assigned to the nearest available network-owned MR (see line 10 of Algorithm 3) or the earliest available one, if there is not any (see line 12 of Algorithm 3). Note that our context-aware scheme gives priority to user-owned MRs, thus substantially reducing OPEX. It is worthwhile to mention that we aim at minimizing the energy consumption of the assigned MR by using our proposed self-aware motion planning (see line 19 of Algorithm 3), to be elaborated on in technically greater detail in Section 3.4.

Next, we present a complexity analysis of our proposed CADMRTC algorithm. We note that the best and worst case time complexity of our proposed algorithm are $\mathcal{O}(|\mathcal{R}_U^A| + n)$ and $\mathcal{O}(|\mathcal{R}_U^A| + |\mathcal{R}_U^B| + |\mathcal{R}_N^A| + |\mathcal{R}_N^B| + n)$, respectively, where n is the number of operations performed by the self-aware optimal robot motion planning (SAOMP) algorithm (see Algorithm 4). We note that n is a constant number that depends on the number of local parameters of a given robot and does not scale with growing numbers of MR. This suggests that the total time complexity of our CADMRTC algorithm is $\mathcal{O}(|\mathcal{R}_U^A| + |\mathcal{R}_U^B| + |\mathcal{R}_N^A| + |\mathcal{R}_N^B|)$.

Algorithm 3 CADMRTC Algorithm

Input: $J_i, t_i^a, \mathcal{R}_U^A, \mathcal{R}_U^B, \mathcal{R}_N^A, \mathcal{R}_N^B, \mathcal{R}, l_j^r, t_j^{av}, D$

Output: $X_i^j, t_j^{av}, l_j^r, \forall j = 1, 2, \ldots, N$

1: **if** $\mathcal{R}_U^A \neq \emptyset$ **then**

2:     $j^* \leftarrow \underset{j \in \mathcal{R}_U^A}{\operatorname{argmin}} \, d\left(l_j^r, l_i^{task}\right)$

3: **else**

4:     **if** $\mathcal{R}_U^B \neq \emptyset$ **then**

5:       $W_{min} \leftarrow \min_{j \in \mathcal{R}_U^B} (t_j^{av} - t_i^a)$

6:       **if** $W_{min} < D$ **then**

7:         $j^* \leftarrow \underset{j \in \mathcal{R}_U^B}{\operatorname{argmin}} (t_j^{av} - t_i^a)$

8:       **else**

9:         **if** $\mathcal{R}_N^A \neq \emptyset$ **then**

10:           $j^* \leftarrow \underset{j \in \mathcal{R}_N^A}{\operatorname{argmin}} \, d\left(l_j^r, l_i^{task}\right)$

11:         **else**

12:           $j^* \leftarrow \underset{j \in S}{\operatorname{argmin}} (t_j^{av} - t_i^a)$

13:         **end if**

14:       **end if**

15:     **end if**

16: **end if**

17: $X_i^{j^*} \leftarrow 1$

18: $\Delta d \leftarrow d\left(l_{j^*}^r, l_i^{task}\right)$

19: $(T_{trav}^{i,j^*}, E_{trav}^{i,j^*}) = \text{SAOMP}\,(\Delta d, j^*)$ (call Algorithm 4)

20: $t_{j^*}^{av} \leftarrow t_{j^*}^{av} + T_{trav}^{i,j^*} + T_{exc}^{i,j^*}$

21: **return** $X_i^j, t_j^{av}, l_j^r, \forall j = 1, 2, \ldots, N$

3.4 Self-Aware Optimal Motion Planning

Battery-powered MRs typically operate for long periods of time. Therefore, it is necessary to optimize their motion by minimizing not only their traverse time but also their energy consumption. In this section, we aim to find the energy-optimal velocity profile of an MR for a given path to traverse.

So far we have derived traverse time T_{trav} of an MR for a given distance Δd and velocity profile $v(t)$. We have shown that an increasing ω_d decreases traverse time T_{trav}. Moreover, our derived closed-form formula for energy consumption of the MR demonstrates that under certain conditions, there exists an $\omega_d^* \in (0, 1)$ for which the energy consumption is minimized. Otherwise, the energy consumption increases for increasing ω_d. Nevertheless, we note that the choice of ω_d is constrained by the maximum achievable acceleration, which in turn depends on the physical design of the motor deployed at the MR. Hence, a_{acc} is given by

$$a_{acc} = -a_{dec} = \frac{v_{max}}{T_{acc}} = \frac{v_{max}}{(1 - \omega_d)\frac{\Delta d}{v_{max}}} \tag{3.17}$$

which implies that a_{acc} is a monotonically increasing function of ω_d with lower and upper bounds given by $\frac{v_{max}^2}{\Delta_d}$ and ∞, which are reached for $\omega_d \to 0$ and $\omega_d \to 1$, respectively. Thus, for a given maximum achievable acceleration a_{acc}^{max}, the feasible range for ω_d is obtained as

$$\frac{v_{max}}{(1 - \omega_d)\frac{\Delta d}{v_{max}}} \leq a_{acc}^{max} \Leftrightarrow \omega_d \leq \overbrace{1 - \frac{v_{max}^2}{a_{acc}^{max} \Delta d}}^{\omega_d^m} \tag{3.18}$$

Next, we aim to minimize both traverse time T_{trav} and $E(\omega_d)$, which under certain conditions may become conflicting objectives. Therefore, to find a compromise between these two conflicting objectives, we aim to solve the following multiobjective optimization problem:

$$\min_{\omega_d} \quad f(\omega_d) = \frac{E(\omega_d)}{E_m} + \frac{T_{trav}(\omega_d)}{T_{trav}^U} \tag{3.19a}$$

$$\text{s.t.} \quad \omega_d \leq \omega_d^m \tag{3.19b}$$

$$0 \leq \omega_d < 1 \tag{3.19c}$$

where T_{trav}^U is given in Eq. (3.6a) and E_m, the upper bound of $E(\omega_d)$, is equal to

$$E_m = \max\{E(0), E(\omega_d^m)\} \tag{3.20}$$

We note that $f(\omega_d)$ is a convex function of ω_d, as it is the sum of two convex functions. For now, we relax the constraint 3.19b and then solve the relaxed optimization problem by letting $\frac{\partial f(\omega_d)}{\partial \omega_d} = 0$ for the following two cases.

Case 1: In this case, $E(\omega_d)$ does not have a local minimum for $\omega_d \in (0, 1)$, i.e. $(\Delta d, v_{max}) \notin \mathbf{A}_1$ in Figure 3.4. Since $E(\omega_d)$ is a monotonically increasing function of ω_d, its upper bound, E_m, is obtained for $\omega_d = 1 - \frac{v_{max}^2}{a_{acc}^{max} \Delta d}$. Thus, we have

$$E_m = 2c_1 v_{max} a_{acc}^{max} + \frac{c_2}{3}\left(3v_{max}\Delta d - \frac{v_{max}^3}{a_{acc}^{max}}\right)$$
$$+ c_3\Delta d + \left(1 + \frac{v_{max}^2}{a_{acc}^{max}\Delta d}\right)\frac{c_4\Delta d}{v_{max}} \tag{3.21}$$

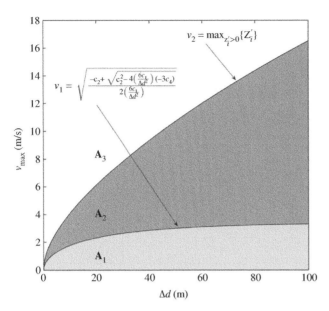

Figure 3.4 Different mobile robot (MR) operational regions represented by A_1, A_2, and A_3 on $\Delta d - v_{max}$ plane, which the proposed self-aware optimal motion planning strategy relies on ($a_{acc}^{max} = 2$ m/s² fixed). Source: Ebrahimzadeh et al. 2019. © 2019 IEEE.

By substituting Eqs. (3.21) and (3.6a) into Eq. (5.28)a, we obtain $f(\omega_d)$ as

$$f(\omega_d) = \frac{1}{E_m}E(\omega_d) + \frac{2 - \omega_d}{2} \tag{3.22}$$

Let ω_d^* denote the optimal value of $\omega_d \in (0, 1)$, for which $f(\omega_d)$ is minimized. We then obtain ω_d^* by solving $\frac{\partial f(\omega_d)}{\partial \omega_d} = 0$, where $\frac{\partial f(\omega_d)}{\partial \omega_d}$ is given by

$$\frac{\partial f(\omega_d)}{\partial \omega_d} = \frac{1}{E_m}\frac{\partial E(\omega_d)}{\partial \omega_d} - \frac{1}{2} \tag{3.23}$$

By substituting Eq. (A.6) into Eq. (3.23), we obtain

$$\frac{\partial f(\omega_d)}{\partial \omega_d} = \frac{v_{max}c_2\Delta d}{3E_m} + \frac{2c_1 v_{max}^3}{\Delta d E_m(1 - \omega_d)^2} - \frac{c_4\Delta d}{v_{max}E_m} - \frac{1}{2} \tag{3.24}$$

Solving $\frac{\partial f(\omega_d)}{\partial \omega_d} = 0$ gives us ω_d^* as

$$\omega_d^* = 1 \pm \sqrt{K'} \tag{3.25}$$

where

$$K' = \frac{2c_1 v_{max}^3}{\Delta d\left(\dfrac{\Delta d c_4}{v_{max}} - \dfrac{\Delta d v_{max}c_2}{3} + \dfrac{E_m}{2}\right)} \tag{3.26}$$

We note that for $K' > 0$ we have $1 + \sqrt{K'} \notin (0, 1)$ and thus it is not acceptable. Whereas $1 - \sqrt{K'}$ lies in interval $(0, 1)$ for a particular range of v_{max}, as specified in the following lemma.

Lemma 3.3 ω_d^* lies in interval $(0, 1)$ if and only if the following inequality holds:

$$v_{max} < \overbrace{\max_{Z_i' > 0: \Im m[Z_i'] = 0} \{Z_i'\}}^{v_2} \tag{3.27}$$

where $\{Z_i'\}_{i=1}^4$ are the roots of the quartic equation given by

$$(A_1 \Delta d - 2c_1) v_{max}^4 + (\Delta d B_1) v_{max}^2 + \Delta d C_1 v_{max} + \Delta d D_1 = 0 \tag{3.28}$$

where

$$A_1 = -\frac{c_2}{3 a_{acc}^{max}}$$

$$B_1 = \left(\frac{2c_2}{3} + c_1 a_{acc}^{max} + \frac{c_4}{a_{acc}^{max}} \right)$$

$$C_1 = \frac{c_3 \Delta d}{2}$$

$$D_1 = \frac{3 \Delta d c_4}{2} \tag{3.29}$$

Proof: See Appendix A.3.

We conclude that for $v_1 < v_{max} < v_2$ (i.e. $(\Delta d, v_{max}) \in \mathbf{A}_2$ shown in Figure 3.4), the optimal value of optimization problem (3.19) is obtained as

$$f^* = \begin{cases} f(\omega_d^*) = f(1 - \sqrt{K'}), & 0 < \omega_d < \omega_d^m \\ f(\omega_d^c), & \text{otherwise} \end{cases} \tag{3.30}$$

For $v_{max} > v_2$ (i.e. $(\Delta d, v_{max}) \in \mathbf{A}_3$ shown in Figure 3.4), on the other hand, $g(\omega_d) = 0$ does not have any root in interval $(0, 1)$. Since $f(\omega_d) > 0$ and $\frac{\partial^2 f(\omega_d)}{\partial \omega_d^2} > 0$ for $\omega_d \in [0, 1)$, and $\lim_{\omega_d \to 1} g(\omega_d) = +\infty$, the optimal value f^* of optimization problem (3.19) is equal to $f(0)$.

Case 2: In this case, which is illustrated by $(\Delta d, v_{max}) \in \mathbf{A}_1$ in Figure 3.4, both $f(\omega_d)$ and $E(\omega_d)$ have a local minimum for $\omega_d \in (0, 1)$. Similarly to Case 1, the optimal value of optimization problem (3.19) is obtained by using Eq. (3.30).

In summary, Algorithm 4 shows the pseudocode of our proposed SAOMP algorithm, which runs locally in the MRs. Given the local parameters of v_{max}, Δd, a_{acc}^{max}, c_1, c_2, c_3, and c_4 of the assigned MR, our proposed self-aware algorithm makes a trade-off between the traversing time T_{trav} and energy consumption E_{trav} by means of optimally planning its motion.

Algorithm 4 SAOMP Algorithm

Input: $v_{max}, \Delta d, a_{acc}^{max}, c_1, c_2, c_3, c_4$
Output: E_{trav}, T_{trav}
1: Use v_1 and v_2 given by Eq. (A.10) and (3.27), respectively, to determine $\mathbf{A}_1, \mathbf{A}_2$, and $\mathbf{A}_3$
2: $\omega_d^m \leftarrow 1 - \frac{v_{max}^2}{a_{acc}^{max}}$
3: **if** $(\Delta d, v_{max}) \in \mathbf{A}_1 \cup \mathbf{A}_2$ **then**
4: $\quad \omega_d^* \leftarrow 1 - \sqrt{K'}$, where K' is given by Eq. (3.26)
5: $\quad$ **if** $\omega_d^* < \omega_d^m$ **then**
6: $\qquad T_{trav} \leftarrow (2 - \omega_d^*)\frac{\Delta d}{v_{max}}$
7: $\qquad E_{trav} \leftarrow E(\omega_d^*)$ given by Eq. (3.7)
8: $\qquad$ Update the MR velocity profile using Eqs. (3.4) and (3.17)
9: $\quad$ **else**
10: $\qquad T_{trav} \leftarrow (2 - \omega_d^m)\frac{\Delta d}{v_{max}}$
11: $\qquad E_{trav} \leftarrow E(\omega_d^m)$ given by Eq. (3.7)
12: $\qquad$ Update the MR velocity profile using Eqs. (3.4) and (3.17)
13: $\quad$ **end if**
14: **end if**
15: **if** $(\Delta d, v_{max}) \in \mathbf{A}_3$ **then**
16: $\quad T_{trav} \leftarrow \frac{2\Delta d}{v_{max}}$
17: $\quad E_{trav} \leftarrow E(0)$ given by Eq. (3.7)
18: $\quad$ Update the MR velocity profile using Eqs. (3.4) and (3.17)
19: **end if**
20: **return** E_{trav}, T_{trav}

Source: Ebrahimzadeh et al. (2019). © 2019 IEEE.

3.5 Delay and Reliability Analysis

In this section, we develop our analytical framework to calculate the average packet transmission delay as well as the human–robot (HR) connection reliability in FiWi-based Tactile Internet infrastructures. In our analysis, we make the following assumptions:

- *Single-hop WLAN*: MUs and MRs are directly associated with an ONU-AP via a wireless single hop, whereby ONU-MPPs serve as ONU-APs.
- *Task arrival model*: MUs act as service demand points, where tasks arrive at random time instants following a Poisson distribution.
- *Traffic model*: The background traffic rate generated by ONUs with attached fixed (wired) subscribers that are directly connected to the backhaul EPON is set to λ_{ONU}.

3.5.1 Delay Analysis

Recall from Section 3.3 that we estimated the scheduling, traversing, and execution delay components of the total task completion time. In this section, we proceed to develop an analytical framework to estimate the packet transmission related delay components of multirobot task execution over FiWi-based Tactile Internet infrastructures.

We build on the analytical frameworks presented in Beyranvand et al. (2017) and Aurzada et al. (2014). We first define the backhaul downstream traffic intensity ρ^u and ρ^d for a TDM passive optical network (PON) ($\Lambda = 1$) and a WDM PON ($\Lambda > 1$) as

$$\rho^u = \frac{\overline{L}}{\Lambda \cdot c_{\text{PON}}} \sum_{q=1}^{O} \sum_{i=0}^{O} \Gamma_{qi}^{\text{PON}} < 1 \qquad (3.31a)$$

$$\rho^d = \frac{\overline{L}}{\Lambda \cdot c_{\text{PON}}} \sum_{q=0}^{O} \sum_{i=1}^{O} \Gamma_{qi}^{\text{PON}} < 1 \qquad (3.31b)$$

where c_{PON} denotes the PON data rate, O denotes the number of ONUs, and Γ_{qi}^{PON} represents the traffic rate (in packets/second) between PON nodes q and i (with $q = 0$ denoting the OLT).

Similar to Aurzada et al. (2014), upstream delay, D_{PON}^u, and downstream delay, D_{PON}^d, of both TDM and WDM PONs are obtained as

$$D_{\text{PON}}^u = \Phi(\rho^u, \overline{L}, \varsigma^2, c_{\text{PON}}) + \frac{\overline{L}}{c_{\text{PON}}} + 2\tau_{\text{PON}} \frac{2 - \rho^u}{1 - \rho^u} - B^u \qquad (3.32)$$

$$D_{\text{PON}}^d = \Phi(\rho^u, \overline{L}, \varsigma^2, c_{\text{PON}}) + \frac{\overline{L}}{c_{\text{PON}}} + \tau_{\text{PON}} - B^u \qquad (3.33)$$

where τ_{PON} denotes the average propagation delay between ONUs and OLT, $\Phi(\cdot)$ is the average queuing delay of an M/G/1 queue characterized by the Pollaczek–Khintchine formula as

$$\Phi(\rho, \overline{L}, \varsigma^2, c) = \frac{\rho}{2c(1 - \rho)} \left(\frac{\varsigma^2}{\overline{L}} + \overline{L} \right) \qquad (3.34)$$

and

$$B^d = B^u = \Phi\left(\frac{\overline{L}}{\Lambda \cdot c_{\text{PON}}} \sum_{q=1}^{O} \sum_{i=1}^{O} \Gamma_{qi}^{\text{PON}}, \overline{L}, \varsigma^2, c_{\text{PON}} \right) \qquad (3.35)$$

Next, we calculate the average delay experienced by an arriving packet at wireless subscribers. Let $D_{z,i}^{e2e}$ denote the average packet delay of wireless subscriber i that resides within the coverage area of ONU-AP$_z$. The set of MUs, $\mathcal{U}_z^{\text{MU}}$, and MRs, $\mathcal{U}_z^{\text{MR}}$, along with their associated ONU-AP$_z$ constitute $\mathcal{U}_z = \mathcal{U}_z^{\text{MU}} \cup \mathcal{U}_z^{\text{MR}} \cup \{0\}$ with $i = 0$ representing ONU-AP$_z$. We then obtain $D_{z,i}^{e2e}$ as

$$D_{z,i}^{e2e} = \frac{1}{\dfrac{1}{\Delta_{z,i}} - \sigma_{z,i}}, \qquad \Delta_{z,i}\sigma_{z,i} < 1, \quad \forall i \in \mathcal{U}_z, \ z = 1, 2, \dots, N_{\text{AP}} \qquad (3.36)$$

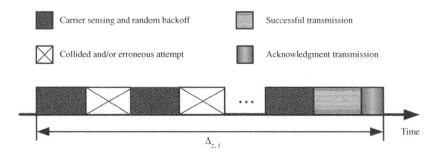

Figure 3.5 Delay components of average channel access delay in IEEE 802.11 distributed coordination function (DCF) with random back-offs. Source: Ebrahimzadeh et al. 2019. © 2019 IEEE.

where $\Delta_{z,i}$ and $\sigma_{z,i}$ denote the average channel access delay and traffic rate, respectively, and N_{AP} is the total number of ONU-APs. Note that Eq. (3.36) accounts for both queuing delay as well as channel access (service) delay of wireless subscriber $i \in \mathcal{U}_z$. We also note that the average access delay $\Delta_{z,i}$ consists of time delays due to carrier sensing, exponential back-offs, collided and erroneous (if any) attempts, successful transmission, and acknowledgement transmission, as illustrated in Figure 3.5.

To compute the average channel access delay, we define a two-dimensional Markov process $(s(t), b(t))$ under unsaturated conditions (see Figure 2.12) and estimate the average service time $\Delta_{z,i}$ in a WLAN using IEEE 802.11 distributed coordination function (DCF) for access control, whereby $b(t)$ and $s(t)$ denote the random back-off counter and size of the contention window at time t, respectively. We note that $\Delta_{z,i}$ is obtained as

$$
\Delta_{z,i} = \sum_{k=0}^{\infty} p_{e,i}^k (1 - p_{e,i}) \left[\sum_{j=0}^{\infty} p_{c,i}^j (1 - p_{c,i}) \right.
$$
$$
\left. \times \left(\left(\sum_{b=0}^{k+j} \frac{2^{\min(b,m)} W_0 - 1}{2} E_s \right) + jT_{c,i} + kT_{e,i} + T_{s,i} \right) \right],
$$
$$
\forall i \in \mathcal{U}_z, z = 1, 2, \dots, N_{AP} \qquad (3.37)
$$

In the following, we proceed to evaluate transmission delay T_{trs}^{dmd} from a given MU to the OLT, transmission delay T_{trs}^{alc} from the OLT to the allocated MR, and transmission delay T_{trs}^o from the MR to the MU.

3.5.1.1 Transmission Delay from MU to OLT

The routing path of an allocation demand transmitted by an MU consists of a single wireless hop and subsequent upstream transmission across the backhaul EPON. Therefore, the average packet transmission delay T_{trs}^{dmd} of an MU to the OLT is estimated as

$$T_{trs}^{dmd} = \underbrace{\mathbb{E}(D_{z,i}^{e2e})}_{\text{MU to ONU-AP}} + \underbrace{D_{PON}^{u}}_{\text{ONU-AP to OLT}} \tag{3.38}$$

where $\mathbb{E}(D_{z,i}^{e2e})$ is computed for $\forall i \in U_z^{MU}, z = 1, 2, \ldots, N_{AP}$.

3.5.1.2 Transmission Delay from OLT to MR

After scheduling, the task coordinator collocated at the OLT transmits the task allocation to the selected MR. Therefore, the average transmission delay T_{trs}^{alc} from the OLT to an MR is given by

$$T_{trs}^{alc} = \underbrace{D_{PON}^{d}}_{\text{OLT to ONU-AP}} + \underbrace{\mathbb{E}(D_{z,i}^{e2e})}_{\text{ONU-AP to MR}}, \tag{3.39}$$

where $\mathbb{E}(D_{z,i}^{e2e})$ is computed for $\forall i = 0, z = 1, 2, \ldots, N_{AP}$.

3.5.1.3 End-to-End Delay from MR to MU

After successfully accomplishing the task, the MR transmits the task output/result to the corresponding MU via the associated ONU-AP. The average transmission delay T_{trs}^{o} from an MR to an MU is then obtained as

$$T_{trs}^{alc} = \underbrace{\mathbb{E}(D_{z,i}^{e2e})}_{\text{MR to ONU AP}} + \underbrace{\mathbb{E}(D_{z,i}^{e2e})}_{\text{ONU AP to MU}} \tag{3.40}$$

where the first term is averaged over $\forall i \in U_z^{MR}, z = 1, 2, \ldots, N_{AP}$, whereas the second term is averaged over $\forall i = 0, z = 1, 2, \ldots, N_{AP}$.

3.5.2 Reliability Analysis

Recall from above that according to recent real-world smartphone traces, WiFi connection and interconnection times follow a truncated Pareto distribution (Beyranvand et al., 2017). The stationary probability that an MU/MR temporarily resides within the coverage area of an ONU-AP is given by

$$P_{temp} = \frac{\overline{T}_{on}}{\overline{T}_{on} + \overline{T}_{off}} = \frac{\overline{T}_{on}/\overline{T}_{off}}{1 + \overline{T}_{on}/\overline{T}_{off}} \tag{3.41}$$

In order for an MR to successfully perform the task requested by an MU, both MU and MR have to be connected to the associated ONU-AP. Let us define the HR connectivity probability P_{HR} as the probability that both MU and MR are connected to the associated ONU-APs, which is given by

$$P_{HR} = P_{temp}^{MU} \cdot (1 - P_{MU \to AP}^{drop}) \cdot (1 - P_{AP \to MR}^{drop}) \cdot P_{temp}^{MR} \cdot (1 - P_{MR \to AP}^{drop}) \cdot (1 - P_{AP \to MU}^{drop}) \tag{3.42}$$

where P^{drop} denotes the packet dropping probability.

Furthermore, let us define the HR connection reliability function, $R_{HR}(t)$, as the probability that the HR connection time lasts longer than t seconds. First, let random variables T_{MU} and T_{MR} denote the WiFi connection lifetime of the MU and MR, respectively. Recall that according to our mobility model, T_{MU} and T_{MR} follow a truncated Pareto distribution, whose probability distribution function (PDF) is given by

$$f(t) = \frac{\alpha \gamma^\alpha}{1 - \left(\frac{\gamma}{v}\right)^\alpha} t^{-(\alpha+1)}, \quad 0 < \gamma \le t \le v \tag{3.43}$$

For notational convenience, we use subscripts H and R to denote the MU and MR, respectively. Note that HR connection time T_{HR} can be computed as

$$T_{HR} = \min\{T_{MU}, T_{MR}\} \tag{3.44}$$

The probability that the HR connection time T_{HR} is greater than t translates to the joint probability

$$P(T_{MU} > t, T_{MR} > t)$$

which in turn gives HR connection reliability $R_{HR}(t)$ as follows:

$$R_{HR}(t) = P(T_{MU} > t) \cdot P(T_{MR} > t) \tag{3.45}$$

We note that this is due to the fact that T_{MU} and T_{MR} are two independent random variables since the mobility of an MU does not depend on that of an MR, thus rendering the MU-AP and MR-AP connections completely independent from each other. Consequently, Eq. (3.45) is then equal to

$$R_{HR}(t) \quad = \left(1 - \underbrace{\int_0^t f_H(t)dt}_{P(T_H)<t}\right)\left(1 - \underbrace{\int_0^t f_R(t)dt}_{P(T_R)<t}\right)$$

$$\underset{=}{\overset{Eq.(3.43)}{=}} \left(1 - \frac{1 - \left(\frac{\gamma_H}{t}\right)^{\alpha_H}}{1 - \left(\frac{\gamma_H}{v_H}\right)^{\alpha_H}}\right)\left(1 - \frac{1 - \left(\frac{\gamma_R}{t}\right)^{\alpha_R}}{1 - \left(\frac{\gamma_R}{v_R}\right)^{\alpha_R}}\right) \tag{3.46}$$

Next, we proceed to estimate the conditional probability of an HR connection failure during time interval $[t, t + \xi]$, given that MU and MR have been connected for the last t seconds:

$$P(t < T_{HR} < t + \xi \mid T_{HR} > t) = \frac{P(\text{connection failure happens in } [t, t + \xi])}{P(T_{HR} > t)} \tag{3.47}$$

As $\xi \to 0$, Eq. (3.47) reduces to

$$P(t < T_{\mathrm{HR}} < t + \xi \mid T_{\mathrm{HR}} > t) = \frac{dF_{T_{\mathrm{HR}}}(t)}{1 - F_{T_{\mathrm{HR}}}(t)} \tag{3.48}$$

The right-hand side of Eq. (3.48), which is commonly referred to as *failure rate function (FRF)* denoted by $h_{\mathrm{HR}}(t)$, represents the conditional probability intensity that an HR connection fails, given that it has lasted up to time t (Ross, 2014). Hence, $h_{\mathrm{HR}}(t)$ is then obtained as

$$h_{\mathrm{HR}}(t) = \frac{\frac{\partial}{\partial t}(1 - R_{\mathrm{HR}}(t))}{R_{\mathrm{HR}}(t)} \tag{3.49}$$

Substituting Eq. (3.46) into Eq. (3.49) and then differentiating with respect to t finally yields

$$h_{\mathrm{HR}}(t) = \frac{\dfrac{\alpha_H \gamma_H^{\alpha_H}}{1 - \left(\dfrac{\gamma_H}{v_H}\right)^{\alpha_H}} t^{-(\alpha_H + 1)}}{\dfrac{1 - \left(\dfrac{\gamma_H}{t}\right)^{\alpha_H}}{1 - \left(\dfrac{\gamma_H}{v_H}\right)^{\alpha_H}}} + \frac{\dfrac{\alpha_R \gamma_R^{\alpha_R}}{1 - \left(\dfrac{\gamma_R}{v_R}\right)^{\alpha_R}} t^{-(\alpha_R + 1)}}{\dfrac{1 - \left(\dfrac{\gamma_R}{t}\right)^{\alpha_R}}{1 - \left(\dfrac{\gamma_R}{v_R}\right)^{\alpha_R}}} \tag{3.50}$$

Note that $F_{T_{\mathrm{HR}}}(t)$ is an increasing failure rate (IFR)/decreasing failure rate (DFR) distribution, if $h_{\mathrm{HR}}(t)$ is an increasing/decreasing function of t. We can easily verify that $h_{\mathrm{HR}}(t)$ in Eq. (3.50) is a convex function of t. Thus, there exists a $t^* > 0$ such that $\frac{\partial}{\partial t} h_{\mathrm{HR}}(t)|_{t^*} = 0$, whereby $\frac{\partial}{\partial t} h_{\mathrm{HR}}(t)$ is negative for $t < t^*$ (i.e. DFR) and positive for $t > t^*$ (i.e. IFR). We note that IFR renders an intuitive concept in that the probability of an HR connection failure increases over time. DFR, on the other hand, implies that the probability of losing an HR connection decreases over time, which happens for $t < t^*$.

3.6 Results

In this section, we investigate the performance of our proposed task allocation scheme. For convenience, we summarized the key parameters and their assigned default values in Table 3.1, which lists the parameter values of the considered FiWi network taken from Beyranvand et al. (2017), those of the MRs in compliance with Tokekar et al. (2014), and those of the physical and digital tasks in consistency with You et al. (2017) and Chowdhury et al. (2018).[4] We consider four ONU-APs as

---

4 It is worthwhile to mention that these parameters are either based on real-world experiments/measurements or in compliance with well-known standards (e.g. IEEE 802.11n/ac and IEEE 802.3ah).

Table 3.1 MR and FiWi network parameters and default values.

Parameter	Value	Parameter	Value
c_1	17.75	DIFS	34 μs
c_2	1.16	SIFS	16 μs
c_3	10.46	PHY header	20 μs
c_4	4.70	W_0	16 slots
γ_0	[0, 25, 50, 75,100]%	H	6
v_{max}	2 m/s	ϵ	9 μs
a_{max}^{acc}	2 m/s²	RTS	20 bytes
$\mathbb{E}(t_i^a)$	8 s	CTS	14 bytes
W_i^p	1–10 kJ	ACK	14 bytes
W_i^d	100–500 MHz (cycles/s)	r in WMN	300 Mbps
N	4	c_{PON} in PON	10 Gbps
N_{MU}	8	l_{PON}	20 km
C^p	1 kW	$\overline{L}$	1500 bytes
C^d	100–500 GHz (cycles/s)	ς_L^2	0
ϵ_d	10×10^{-10} J	ONU-AP radius	$10\sqrt{2}$ m

Source: Ebrahimzadeh et al. 2019. © 2019 IEEE.

well as four ONUs serving fixed wired subscribers. Associated with each ONU-AP are two MUs along with an MR, with a total of eight MUs and four MRs. We use a Poisson point process to generate the random locations of MUs and MRs in an 80×80 m² area. We assume that task demands arrive at the task coordinator with exponential interarrival times.

Figure 3.6 depicts the average OPEX per task vs. user- to network-ownership cost ratio, $r_{U2N} = \frac{\varphi_U}{\varphi_N} \leq 1$. Interestingly, we observe that while full user-ownership (i.e. $\gamma_0 = 100\%$) is always beneficial for MUs in terms of OPEX savings, partial user-ownership (i.e. $\gamma_0 < 100\%$) isn't necessarily so. From Figure 3.6 we observe that a user-ownership of $\gamma_0 = 25\%$ is less costly than full network-ownership (i.e. $\gamma_0 = 0\%$) only for $r_{U2N} < 0.39$. For $r_{U2N} > 0.39$, on the other hand, MUs face lower OPEX per task with full network-ownership ($\gamma_0 = 0\%$) compared with a partial user-ownership of $\gamma_0 = 25\%$. The reason for this is that for $\gamma_0 = 0\%$, our task coordination algorithm allocates tasks to the preferred user-owned MR(s), which is (are) responsible for giving service to all MUs in the area. This, in turn, increases the average distance traversed by user-owned MRs, thus increasing average traverse time and energy consumption. As $\frac{\varphi_U}{\varphi_N}$ becomes greater than 0.39, full network-ownership therefore proves less costly. Further, note that in order

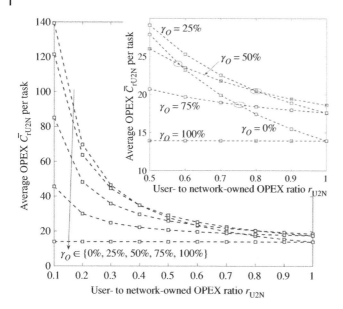

Figure 3.6 Average cost, $\overline{C}$, per executed task vs. user- to network-owned operational expenditures (OPEX) ratio r_{U2N} ($D = 0$ fixed). Source: Ebrahimzadeh et al. 2019. © 2019 IEEE.

for a partial user-ownership of $\gamma_O = 50\%$ and $\gamma_O = 75\%$ to be less costly than full network-ownership ($\gamma_O = 0\%$), r_{U2N} must not exceed 0.58 and 0.73, respectively (see also Figure 3.6). Moreover, we observe that as $\frac{\varphi_U}{\varphi_N} \to 0$, the beneficial impact of user-ownership on OPEX savings is more pronounced, whereas for $\frac{\varphi_U}{\varphi_N} \to 1$, the average OPEX $\overline{C}_{r_{U2N}}$ per task for different values of γ_O converges to that of $\gamma_O = 100\%$. This is because as $\frac{\varphi_U}{\varphi_N} \to 1$ we have $\varphi_U \approx \varphi_N$, thus user-ownership does not reveal a notable OPEX gain compared with full network-ownership.

Next, we explore the impact of increasing waiting deadline[5] D on OPEX savings in Figure 3.7, which depicts the average OPEX, $\overline{C}$ per executed task, vs. waiting deadline D. We find that an increasing D reduces $\overline{C}$ only for partial user-ownership (i.e. $\gamma_O = 25\%$, 50%, and 75%). To better understand this, let θ denote the ratio of the number of executed tasks by user-owned MRs to the total number of tasks. Figure 3.8 depicts θ vs. waiting deadline D for the same fixed $r_{U2N} = 0.2$. We observe that an increasing D has no impact on $\gamma_O = 0\%$ and 100%, whereas it increases θ for $\gamma_O = 25\%$, 50%, and 75%. This is due to the fact that

---

5 We note that by setting $D = 0$ for $\gamma_O = 0\%$ and 100%, our proposed CADMRTC algorithm may be viewed as the nearest available robot allocation scheme, which stands as a baseline for fair comparison.

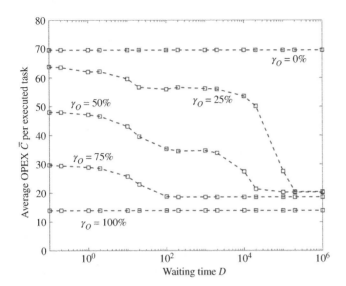

Figure 3.7 Average operational expenditures (OPEX), $\bar{C}$, per executed task vs. waiting deadline D ($r_{U2N} = 0.2$ fixed). Source: Ebrahimzadeh et al. 2019. © 2019 IEEE.

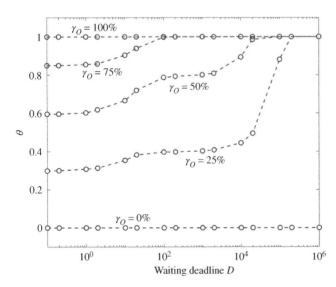

Figure 3.8 θ vs. waiting deadline D ($r_{U2N} = 0.2$ fixed). Source: Ebrahimzadeh et al. 2019. © 2019 IEEE.

for increasing D more tasks are executed by user-owned MRs rather than their network-owned counterparts, resulting in a decreased $\overline{C}$ (see also Figure 3.7).

Next, we plot the average task completion time vs. waiting deadline in Figure 3.9. The figure shows that for γ_O the average task completion time increases linearly with D. We note that Figure 3.9 along with Figures 3.7 and 3.8 demonstrate that for $\gamma_O = 0\%$, setting $D = 0$ achieves the best performance in terms of not only OPEX but also average task completion time. For $\gamma_O = 25\%$, 50%, and 75%, on the other hand, the average task completion time increases for increasing D until it hits a plateau. The values of D above, with the average task completion time remaining constant, are obtained as 10^5, 2×10^4, and 100 seconds for $\gamma_O = 25\%$, 50%, and 75%, respectively.

The obtained 2-D Pareto front results of our proposed CADMRTC algorithm are depicted in Figure 3.10, which characterizes the trade-off between the average OPEX per task and average task completion time. Figure 3.10 reveals that none of the obtained results for a given γ_O is dominant, thus the decision-maker can yield a flexible trade-off between the two objectives of the problem by appropriately setting the waiting deadline D. Figure 3.11 depicts the average task completion time vs. ownership spreading factor γ_O for different deadline $D \in \{0, 2, 5, 10\}$ (given in seconds). Generally, we observe a trend of decreasing average task completion time for increasing ownership spreading factor, whereby the impact of varying D becomes negligible for an ownership spreading factor of 75% and higher. Note that the lowest average task completion time of roughly 16 seconds can be achieved for

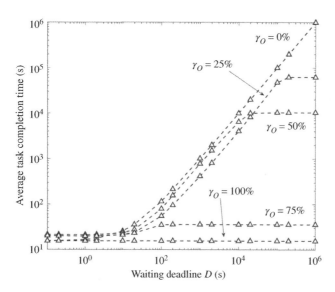

Figure 3.9 Average task completion time vs. waiting deadline D. Source: Ebrahimzadeh et al. 2019. © 2019 IEEE.

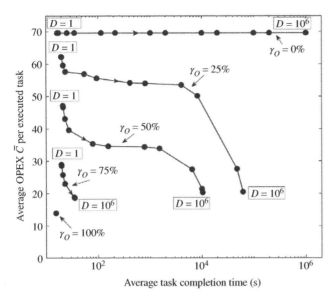

Figure 3.10 2-D Pareto-front of our proposed context-aware dynamic multirobot task coordination (CADMRTC) algorithm for different values of ownership spreading factor γ_O (waiting deadline D increases along the arrow shown on each curve). Source: Ebrahimzadeh et al. 2019. © 2019 IEEE.

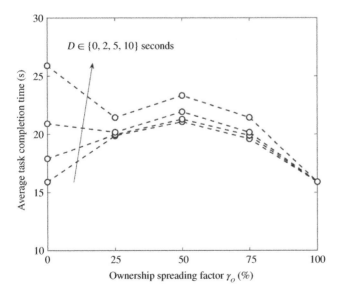

Figure 3.11 Average task completion time vs. ownership spreading factor. source Source: Ebrahimzadeh et al. 2019. © 2019 IEEE.

$D = 0$ with either 0% or 100% ownership spreading. This is due to the fact that in both cases all MRs are eligible for immediate task allocation. More interestingly, for $D = 0$ and to a lesser extent also for $D = 2$ seconds the average task completion time increases for an ownership spreading factor of up to 50%, as opposed to the aforementioned general trend. This observation stems from the unbalanced task allocation between a few overutilized user-owned MRs and the rest of underutilized network-owned MRs (see also Figure 3.8).

Figure. 3.12 illustrates the probability of HR connectivity vs. $\overline{T}_{on}^{MU} / \overline{T}_{off}^{MU}$ for different values of $\overline{T}_{on}^{MR} / \overline{T}_{off}^{MR}$. For $\overline{T}_{on} / \overline{T}_{off} = 2.73$, which is obtained from real-world measurements in Beyranvand et al. (2017), we achieve a maximum of 53.57% HR connectivity probability. Note that for an increasing $\overline{T}_{on}^{MU} / \overline{T}_{off}^{MU}$ of up to 30, P_{HR} increases until it levels off. Conversely, for $\overline{T}_{on}^{MU} / \overline{T}_{off}^{MU} > 30$, P_{HR} highly depends on the temporal availability of MR, P_{temp}^{MR}.

Finally, Figure 3.13 shows the HR connection reliability function $R_{HR}(t)$ and HR connection failure rate $h_{HR}(t)$. Note that for $t < 615.4$ minutes, the reliability function is DFR, whereas it is IFR for $t > 615.4$. This implies that the connection failure rate $h_{HR}(t)$ decreases as time t increases up to $t^* = 615$ minutes, given that it has not failed by time t. At $t = t^*$, the minimum value of 0.0044 is achieved. For $t > t^*$, on the other hand, the HR connection failure rate increases up to roughly 120 min^{-1} (i.e. average interfailure time becomes 500 ms).

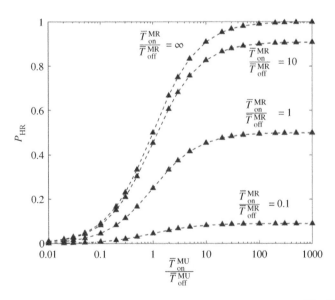

Figure 3.12 Human–robot (HR) connectivity probability vs. $\overline{T}_{on}^{MU} / \overline{T}_{off}^{MU}$ for different values of $\overline{T}_{on}^{MR} / \overline{T}_{off}^{MR}$. Source: Ebrahimzadeh et al. 2019. © 2019 IEEE.

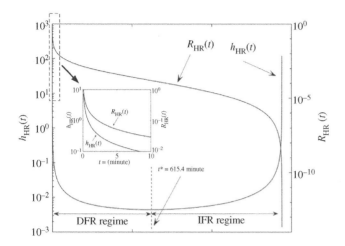

Figure 3.13 Human–robot (HR) connection reliability function $R_{HR}(t)$ and failure rate function $h_{HR(t)}$ vs. time. Source: Ebrahimzadeh et al. 2019. © 2019 IEEE.

3.7 Conclusion

We investigated the performance of our proposed context- and self-aware HART centric multirobot task allocation over FiWi-based Tactile Internet infrastructures. We shed light on when, how, and under which circumstances user-ownership of MRs becomes beneficial in terms of OPEX per executed task. Further, we evaluated the performance of our proposed CADMRTC algorithm in terms of average task completion time, OPEX per executed task, and ratio of the number of executed tasks by user-owned MRs and the total number of tasks. By leveraging on the low-latency and reliable fiber backhaul and distributed WiFi-based fronthaul, we showed that a HR connectivity probability of > 90% is achievable for $\overline{T}_{on}^{MR} / \overline{T}_{off}^{MR} >$ 10. In addition, our obtained results show that our proposed self-aware scheme plays a key role in minimizing the traverse time as well as energy consumption of MRs in a distributed manner, whereas our context-aware task coordination is instrumental in minimizing the task completion time, while paying particular attention to reducing OPEX of user-/network-ownership of MRs.

Importantly, our obtained results show that from a performance perspective (in terms of average task completion time) almost no deterioration occurs if the ownership is shifted entirely from network operators to mobile users ($D = 0$), though such a shift in ownership of robots has significant implications on sharing the profits and collaborative business opportunities arising from the emerging Tactile Internet in a more equitable fashion. As a result, this may open up new opportunities for synergies between humans and machines/robots, while spurring the symbiotic human–machine/HR development envisaged by earlyday Internet pioneers and imagining entirely new categories of abundance for a low entry cost economy. Among others, one future research direction is to further explore the synergies between the aforementioned HART membership and the complementary strengths of robots to facilitate local human–machine coactivity clusters by decentralizing the Tactile Internet. Another interesting open research problem is how human *crowdsourcing* can help decrease task completion time in the event of unreliable connectivity and/or network failures. Note that our presented spreading ownership of robots across mobile users may be an important stepping stone to collaborative business relationships that function more like localized share-economy ecosystems than markets.

4

Delay-Constrained Teleoperation Task Scheduling and Assignment

4.1 Introduction

A popular misinterpretation about robotics is that intelligent systems, ranging from advanced robots to digital bots, will gradually substitute humans in one job after another. This argument may be true for some jobs, but we note that even though advanced robotics can be deployed to automate certain jobs, its greater potential, yet to be unleashed, is to *complement* and *augment* human capabilities. The cutting-edge jobs and innovative businesses that arise from human–machine symbiosis are happening in the so-called *missing middle* that refers to the new ways that have to bridge the gap between human-only and machine-only activities. This gives way to the so-called *third wave of business transformation*, which will be centered around human + machine hybrid activities (Daugherty and Wilson, 2018). Key toward developing the missing middle is to understand the ways humans help machines and the ways machines help humans. An interesting example of recognizing the relative strengths of humans and machines and leveraging on them to fill the missing middle can be found at automobile manufacturer Audi. Having deployed a fleet of Audi robotic telepresence (ART) systems, Audi has set forth toward employee augmentation that not only helps train technicians in diagnostics and repair but also accelerates delivery of service to customers (Audi and VGo, 2014).

The advent of semiautonomous robotic assistance systems is becoming a part of the vision of the Tactile Internet. An early example is the European research project Robot-Era, which recently concluded the world's largest real-life trial of robot aids for the ageing population. With their small-stage deployment proven successful, robotic helpers will need to request human assistance every now and then, as stated recently by automobile manufacturer Nissan to augment their autonomous vehicle technology with a crew of on-call remote human

Toward 6G: A New Era of Convergence, First Edition. Amin Ebrahimzadeh and Martin Maier.

operators (HOs) acting as "mobility managers," who can remotely take control in unexpected situations (Nowak, 2017).

While the Tactile Internet has been more recently also referred to as the 5G-enabled Tactile Internet, the importance of the so-called *backhaul bottleneck* needs to be recognized as well, calling for an end-to-end design approach leveraging both wireless frontend and wired backhaul technologies (Maier and Ebrahimzadeh, 2019). As mentioned in Chapter 2, this mandatory end-to-end design approach is fully reflected in the key principles of the reference architecture within the IEEE P1918.1 standards working group Aijaz et al. (2018). These key principles aim to develop a generic Tactile Internet reference architecture, supporting local area as well as wide area connectivity through wireless (e.g. cellular, WiFi) or hybrid wireless/wired networking. The importance of such a design approach is more highlighted for Tactile Internet applications that may not always require mobility, e.g. remote healthcare.

Unlike their fully autonomous counterparts, semiautonomous robotic systems rely on human assistance from time to time via teleoperation and/or telepresence when domain expertise is needed to accomplish a specific task, thus allowing for an human-in-the-loop (HITL) centric design approach. As these robots will need to request human assistance via teleoperation/presence, mapping these requests to the HOs themselves stands as a difficult multicriteria optimization problem with the objectives of minimizing the average weighted task completion time, maximum tardiness, and average operational expenditures (OPEX) per task. The difficulty of solving such a problem lies in the following reasons. First, it is clear that we are dealing with different conflicting objectives, which makes it challenging to obtain a satisfactory result, especially for large-sized problem instants. Second, the assignment of a given task to an HO is subject to strict end-to-end packet delay constraints, thus calling for a cross-layer approach, taking into account the delay experienced by packets in both command and feedback paths (to be discussed below).

In this chapter, we formulate and solve the problem of joint prioritized scheduling and assignment of delay-constrained teleoperation tasks to HOs so as to minimize the average weighted task completion time, maximum tardiness, and average OPEX per task. In particular, the contributions of this chapter are as follows:

- We elaborate on the role of FiWi-enhanced networks as the underlying communications infrastructure for enabling emerging delay-sensitive Tactile Internet applications. In particular, trying to build on our findings in Beyranvand et al. (2017) and Maier and Ebrahimzadeh (2019), we aim to realize local and/or non-local teleoperation over FiWi-enhanced networks.
- We define the problem of joint prioritized scheduling and assignment of delay-constrained teleoperation tasks onto available skilled HOs. After

formulating our multiobjective optimization problem, we propose our so-called "context-aware prioritized scheduling and task assignment" (CAPSTA) algorithm to achieve satisfactory results by making suitable trade-offs between the conflicting objectives of the problem.

- We develop our analytical framework to estimate the end-to-end packet delay of both local and nonlocal teleoperation over FiWi-enhanced networks. Our analysis flexibly allows for the coexistence of both conventional human-to-human (H2H) and haptic human-to-machine (H2M) traffic, while focusing on the HOs and teleoperator robots (TORs) involved in either local or nonlocal teleoperation. The results of our delay analysis are then fed into the proposed CAPSTA algorithm.

The remainder of the chapter is structured as follows: Section 4.2 describes FiWi-based Tactile Internet infrastructures for HITL-centric teleoperation-based task coordination. In Section 4.3, we present our problem formulation, which is then solved by proposing our context-aware task coordination algorithm in Section 4.4. Our end-to-end packet delay analysis is presented in Section 4.5. In Section 4.6, we present our obtained results and findings. In Section 4.7, we present a complementary discussion of our findings and point to some interesting future research avenues. Section 4.8 concludes the chapter.

4.2 System Model and Network Architecture

Recall from Chapter 2 that a typical bilateral teleoperation system realizes bidirectional haptic communications between an HO and a TOR, which are both connected via a communication network, as shown in Figure 2.4. In a typical teleoperation system, the position-orientation samples are transmitted from the HO through the human system interface (HSI) in the command path, whereas the force–torque samples are fed back to the HO in the feedback path. By interfacing with the HSI, the HO commands the motion of the TOR in the remote environment. This couples the HO closely with the remote environment and thereby creates a more realistic feeling of remote presence. Following the packetization process presented in Chapter 2, we assume that the haptic packets are of size $8N_{DoF}$ + 40 bytes, accounting for the real-time transport protocol (RTP)/user datagram protocol (UDP)/Internet protocol (IP) header.

Figure 4.1 illustrates the generic network architecture of our considered FiWi-enhanced Long-Term Evolution Advanced (LTE-A) HetNets. As shown in Figure 4.1, selected mobile users (MUs) are equipped with TORs, which are capable of performing physical tasks (simply referred to as tasks hereafter) by establishing haptic communications with HOs. The MUs that are collocated with the TORs act as task demand points. Typically, the number of task demands is

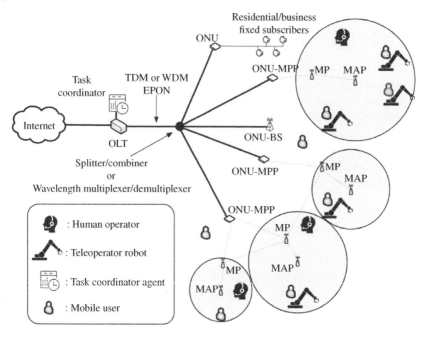

Figure 4.1 Generic architecture of fiber-wireless (FiWi)-based Tactile Internet network infrastructure for teleoperation task coordination. Source: Ebrahimzadeh and Maier (2019). © 2019 IEEE.

greater than that of available skilled HOs. This necessitates a suitable mapping of tasks to the available HOs. Given the set of tasks and available skilled HOs, the task coordinator agent is responsible for the assignment of tasks and scheduling them on the HOs (see Figure 4.1). Note that teleoperation-based tasks arrive at the demand points. The corresponding MUs then send their demands upstream to the task coordinator agent, which is collocated with the optical line terminal (OLT) (see Figure 4.1), via the wireless front-end and Ethernet passive optical network (EPON) backhaul until they reach the OLT. The task coordinator agent then transmits the schedule to the HOs as well as demand points. According to the schedule received from the task coordinator agent, an HO may be involved in either local or nonlocal teleoperation with the corresponding TOR, depending on the proximity of the involved HO and TOR, as illustrated in Figure 4.1. In local teleoperation, the HO and corresponding TOR are associated with the same mesh access point (MAP) and exchange their command and feedback samples through this MAP without traversing the fiber backhaul. Conversely, if HO and TOR are associated with different MAPs, nonlocal teleoperation is generally done by communicating via the backhaul EPON and central OLT.

4.3   Problem Statement

We consider the problem of joint assignment and scheduling of N delay-constrained teleoperation tasks on any fixed number M of HOs as follows. Let $\mathcal{M} = \{O_1, O_2, \dots, O_M\}$ and $\mathcal{J} = \{J_1, J_2, \dots, J_N\}$ denote the set of M available HOs and N given tasks, respectively. Let T_j denote the operation time of task $J_j \in \mathcal{J}$. Note that operation time T_j is given by

$$T_j = s_j + w_j \tag{4.1}$$

where s_j and w_j is the teleoperation session setup time and workload (both in seconds) of task J_j, respectively. Each task $J_j \in \mathcal{J}$ has a due time D_j and is associated with weight Ω_j. Larger weights correspond to higher priority levels. Although the tasks are expected to be accomplished by the given due time, any incurred tardiness is subject to a cost penalty (to be elaborated on in technically greater detail shortly).

We consider an *offline* scheduling scenario, where all tasks are available at time zero and remain available continuously thereafter. Each task can be operated by only one HO at any time and each HO can operate only one task at a time. We also assume that preemption is not allowed, meaning that tasks cannot be split. This is because if tasks were divided and scheduled in noncontinuous time periods, preemption would incur extra reconfiguration/setup overhead, which is significant when the setup time is nonnegligible. For simplicity, we assume, without loss of generality, that operation times, due times, and priority weights are all integers. Further, we assume $N \gg M$. For task J_j, the start and completion times are denoted by S_j and C_j, respectively. A feasible assignment/schedule specifies when and by which HO a given task is operated. Given a feasible schedule, one can compute the tardiness of task J_j as $\max\{0, C_j - D_j\}$. The goal is to assign the tasks to the HOs such that the following constraints are satisfied: (i) no more than one task is assigned to an HO at a time, (ii) no task is assigned to more than one HO, (iii) tasks are not preempted, and (iv) the average end-to-end packet delay of a scheduled teleoperation does not exceed a given delay threshold.

4.3.1   Problem Formulation

We formulate our mixed integer programming (MIP) problem of joint prioritized scheduling and assignment of delay-constrained teleoperation tasks onto HOs as follows:

Given:

- $\mathcal{J}$: Set of tasks.
- $\mathcal{M}$: Set of available HOs.

- J_j: Task $j, j = 1, 2, \ldots, N$.
- T_j: Operation time of task $J_j, j = 1, 2, \ldots, N$.
- Ω_j: Weight of task $J_j, j = 1, 2, \ldots, N$.
- D_j: Due time of task $J_j, j = 1, 2, \ldots, N$.
- O_k: HO $k, k = 1, 2, \ldots, M$.
- $\mathbf{D}_c$: Average end-to-end packet delay matrix of teleoperation pairs in the command path.
- $\mathbf{D}_f$: Average end-to-end packet delay matrix of teleoperation pairs in the feedback path.

Parameters:

- ϵ_h: Operational cost per time unit of tardiness.
- ϵ_m: Operational cost of activating a teleoperation session.
- ϵ_k: Operational cost per time unit of performing a teleoperation task by HO O_k.

Decision variables:

- δ_{ij}: A binary variable, which equals 0 unless task J_i precedes task J_j.
- z_{jk}: A binary variable, which equals 0 unless task J_j is assigned to HO O_k.
- y_{ij}: A binary variable, which equals 0 unless tasks J_i and J_j are not assigned to the same HO.
- S_j: Operation start time associated with task $J_j, j = 1, 2, \ldots, N$.
- C_j: Operation completion time associated with task $J_j, j = 1, 2, \ldots, N$.
- X: Set of total decision variables of the problem represented by $(\{\delta_{ij}\}, \{y_{ij}\}, \{z_{jk}\}, \{S_j\}, \{C_j\})$.

Objective functions:

- $L(X)$: Average weighted task completion time.
- $T(X)$: Maximum tardiness.
- $C(X)$: Operational expenditure.

Multiobjective formulation:

$$\underset{X}{\text{minimize}} \quad L(X), T(X), C(X) \tag{4.2}$$

subject to

$$\delta_{ij} + \delta_{ji} + y_{ij} = 1; \qquad i, j \in \mathcal{J}, i < j \tag{4.3a}$$

$$\delta_{ij} + \delta_{jl} + \delta_{lj} \leq 2; \qquad i, j, l \in \mathcal{J}, i < j < l \tag{4.3b}$$

$$z_{ik} + z_{jk} + y_{ij} \leq 2; \qquad i, j \in \mathcal{J}, i < j, k \in \mathcal{M} \tag{4.3c}$$

$$\sum_{k=1}^{M} z_{jk} = 1; \qquad \forall j \in \mathcal{J} \tag{4.3d}$$

$$C_j \geq T_j z_{jk}; \qquad\qquad j \in \mathcal{J}, i < j, k \in \mathcal{M} \qquad\qquad (4.3e)$$

$$C_j \geq C_i + T_j(\delta_{ij} + z_{ik} + z_{jk} - 2) - K(1 - \delta_{ij}); \qquad\qquad (4.3f)$$

$$i,j \in \mathcal{J}, k \in \mathcal{M}$$

$$\sum_{k \in \mathcal{M}} D^c_{kj} z_{jk} \leq D_0; \qquad\qquad j \in \mathcal{J} \qquad\qquad (4.3g)$$

$$\sum_{k \in \mathcal{M}} D^f_{jk} z_{jk} \leq D_0; \qquad\qquad j \in \mathcal{J} \qquad\qquad (4.3h)$$

$$\delta_{ij}, \delta_{ji}, y_{ij}, z_{jk} \in \{0,1\}; \qquad i,j \in \mathcal{J}, k \in \mathcal{M} \qquad\qquad (4.3i)$$

$$C_j \in \mathbb{R}^+; \qquad\qquad j \in \mathcal{J} \qquad\qquad (4.3j)$$

where $L(X)$, $T(X)$, $C(X)$ are given as follows: Our first objective is to minimize the average weighted task completion time $L(X)$, which is given by

$$L(X) = \frac{1}{N}\sum_{j \in \mathcal{J}} \Omega_j C_j. \qquad\qquad (4.4)$$

The second objective is to minimize the maximum tardiness $T(X)$, which is given by

$$T(X) = \max_{j \in \mathcal{J}} \quad \overbrace{\max\{C_j - D_j, 0\}}^{\text{tardiness of task } j} \qquad\qquad (4.5)$$

which stands as a nonlinear objective function of the decision variables. The third objective is to minimize OPEX, $C(X)$, which is estimated as

$$C(X) = M \cdot \epsilon_m + \sum_{j \in \mathcal{J}} \epsilon_h \Omega_j \max\{C_j - D_j, 0\} + \sum_{k \in \mathcal{M}}\sum_{j \in \mathcal{J}} z_{jk}\epsilon_k T_j \qquad\qquad (4.6)$$

where the first term represents the cost of activating M teleoperation sessions, the second term penalizes the tardy tasks according to their priority levels, i.e. the tardy tasks with higher priorities are subject to higher incurred cost penalty, and the third term models the total cost of performing tasks by HOs. The aforementioned definitions clearly indicate that these objectives are independent and often conflicting optimization targets.

In our aforementioned MIP formulation, constraint set (4.3a) ensures that if tasks J_i and J_j are assigned to the same HO (i.e. $y_{ij} = 0$), one of them should precede the other, thus either δ_{ij} or δ_{ji} must equal 1. On the other hand, if the tasks are assigned to different HOs (i.e. $y_{ij} = 1$), both δ_{ij} and δ_{ji} must equal zero. Constraint (4.3b) ensures a linear ordering of the tasks. According to constraint (4.3c), when tasks J_i and J_j are assigned to HO O_k, then y_{ij} must equal zero. Constraint (4.3d) ensures that each task is assigned to one of the available HOs. Constraints (4.3e) and (4.3g) represent the completion time of the scheduled tasks. We note that in constraint set (4.3g), K is a relatively large number, which is set to $\sum_{j=1}^{N} T_j$ in our problem. Constraints (4.3h) and (4.3i) ensure that the average end-to-end packet delay of any scheduled teleoperation pair HO–TOR in both command and feedback paths is kept below a given threshold D_0. To be more specific, among all the possible HO assignments, the teleoperation pairs that are incurred with an excessive amount of connection latency are excluded from the feasible set.

Note that the average end-to-end packet delays of any possible HO–TOR pair are characterized by two matrices, one of which represents the command path, whereas the other accounts for the feedback path. More specifically, the command delay matrix $\mathbf{D}_c$ is given by

$$
\mathbf{D}_c =
\begin{bmatrix}
D_{11}^c & D_{12}^c & \cdots & D_{1N}^c \\
D_{21}^c & D_{22}^c & \cdots & D_{2N}^c \\
\vdots & & \ddots & \vdots \\
D_{M1}^c & D_{M2}^c & \cdots & D_{MN}^c
\end{bmatrix}_{M \times N}
\tag{4.7}
$$

where element D_{kj}^c in row k ($k = 1, \ldots, M$) and column j ($j = 1, \ldots, N$) denotes the average end-to-end packet delay between HO k and TOR j. Similarly, the feedback delay matrix $\mathbf{D}_f$ is given by

$$
\mathbf{D}_f =
\begin{bmatrix}
D_{11}^f & D_{12}^f & \cdots & D_{1M}^f \\
D_{21}^f & D_{22}^f & \cdots & D_{2M}^f \\
\vdots & & \ddots & \vdots \\
D_{N1}^f & D_{N2}^f & \cdots & D_{NM}^f
\end{bmatrix}_{N \times M}
\tag{4.8}
$$

where element D_{jk}^f in row j ($j = 1, \ldots, N$) and column k ($k = 1, \ldots, M$) denotes the average end-to-end packet delay between TOR j and HO k. Note that the elements of the delay matrices $\mathbf{D}_c$ and $\mathbf{D}_f$, which depend on the state of the underlying network, are estimated by using our delay analysis presented in Section 4.5.

4.3.2 Model Scalability

Recall from above that the problem of assigning and scheduling of tasks to the HOs is subject to strict end-to-end packet delay constraints, which limits the feasible set. If we consider a special case where there are tasks to be mapped to HOs without any end-to-end packet delay constraint and with only one objective of minimizing the average weighted task completion time, then the problem reduces to a parallel machine scheduling problem, which is known to be $\mathcal{NP}$-hard (Brucker, 2007). Given that the single-criterion parallel machine scheduling without any end-to-end delay constraint is a special case of the multicriteria delay-constrained teleoperation task scheduling and assignment, this makes the latter also $\mathcal{NP}$-hard by restriction. Given a set of N tasks and M HOs, the developed formulation has $2N^2 + 2N + N \cdot M$ variables and $\frac{N(N-1)}{2} + \frac{N(N-1)(N-2)}{6} + \frac{2N \cdot M(N-1)}{2} + 2N^2 + N \cdot M + 4N$ constraints, which, along with the conflicting objectives, drastically restrict the scalability of the model even for small-sized problems, therefore calling for algorithmic solutions.

4.4 Algorithmic Solution

4.4.1 Illustrative Case Study

For illustration, we present a case study in order to better understand the impact of different prioritized and nonprioritized coordination strategies on the delay/cost performance from the viewpoint of both users and network operator. Let us consider two HOs and five tasks, as shown in Figure 4.2, where the task parameters (i.e. operation times, due times, and weights[1]) as well as the command/feedback delay matrices (in millisecond) are illustrated.

Strategy **A**, regardless of task weights, assigns the tasks to the nearest HO that resides within the coverage area of the same access point, thus giving preference to realize local teleoperation sessions. Therefore, in strategy **A**, among the feasible solutions that meet the delay constraints specified by Eqs. (4.3g) and (4.3h), tasks J_1, J_2, and J_5 are assigned to O_1, whereas J_3 and J_4 are assigned to O_2. In contrast, Strategy **B** relies on giving preference to high-priority tasks with shorter due

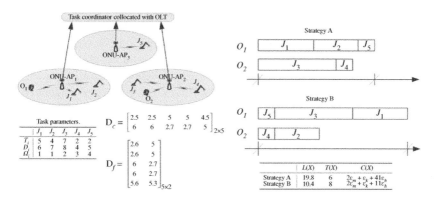

Figure 4.2 An illustrative case study of the delay/cost performance of two different task coordination strategies. Source: Ebrahimzadeh and Maier (2019).© 2019 IEEE.

---

1 Weight is usually related to the importance, while due time is associated with the urgency of a given task and a prioritized scheduler must prepare a sequence able to first perform high-priority tasks.

times, thus J_1, J_3, and J_5 are assigned to O_1, whereas J_2 and $J4$ are assigned to O_2. The results indicate that strategy **B** yields a lower average weighted task completion time and smaller OPEX compared to strategy **A**. We note, however, that such superior performance is achieved at the expense of a 20% increase in maximum tardiness (see also Figure 4.2).

4.4.2 Proposed Task Coordination Algorithm

We note that while the first objective function, $L(X)$, aims to minimize the average weighted task completion time without considering the due times, the second and third objective functions (i.e. $T(X)$ and $C(X)$) deal with the tardiness incurred by overdue completion of tasks, thus they do consider the task due times. Also note that the second objective, $T(X)$, represents the maximum task tardiness, which is preferred to be minimized from a user standpoint. In addition, the third objective, $C(X)$, which addresses the operator revenue, tries to push the task completion times toward minimizing the incurred OPEX, thus implicitly minimizing the average weighted tardiness. This justifies the selection of the three different objectives in our problem formulation.

Clearly, a so-called "optimum" with respect to one objective may perform extremely bad with respect to other criteria (see example in Figure 4.2). Therefore, a nonoptimal solution with satisfactory performance in terms of other measures might be considered a better alternative by the decision-maker. For large-sized problem instants of the developed formulation, the computational difficulties associated with finding a satisfactory solution increase dramatically. Therefore, in order to find a suitable trade-off between the conflicting objectives, we propose our so-called CAPSTA algorithm, which is illustrated in Algorithm 5. The suitable performance of the proposed algorithm relies on an accurate estimation of the context parameters (e.g. task parameters, delay matrices in both command and feedback paths, location of MUs/HOs/TORs, incoming H2H/H2M traffic pattern). In the design of the proposed CAPSTA algorithm, we adopt two sorting policies (to be elaborated on shortly), in both assignment and scheduling phases, in order to perform in favor of high-priority tasks with shorter due times.

As a first step, the proposed CAPSTA algorithm aims to partition the given task set $\mathcal{J}$ into M subsets. Toward this end, our sorting policy indicates that the given tasks are sorted in a decreasing order of $\frac{\Omega_j}{T_j}$ (see line 1 in Algorithm 5). Next, the tasks are selected from the sorted set and then are assigned to the HOs in a round-robin fashion (see lines 3–15 in Algorithm 5). We note, however, that the assignment of task J_n to the HO O_m is valid only if the estimated average end-to-end delays in both command and feedback paths satisfy the delay constraints in Eqs. (4.3g) and (4.3h). Otherwise, we select the HO that corresponds to the minimum average end-to-end delay with task J_n in both

---

Algorithm 5 CAPSTA Algorithm

Input: $\mathcal{J}, \mathcal{M}, T_j, \Omega_j, D_j; \forall j \in \mathcal{J}, \mathbf{D}_c, \mathbf{D}_f$
Output: $S_j, C_j, z_{jk}; \forall j \in \mathcal{J}, \forall k \in \mathcal{M}$

1: Sort $\mathcal{J}$ in a decreasing order of $\frac{\Omega_j}{T_j}, \forall j \in \mathcal{J}$
2: $k \leftarrow 0$
3: **for** $j = 1$ to N **do**
4:    $k \leftarrow k + 1$
5:    $k^* \leftarrow \begin{cases} \mod(k, M) & \text{if } \mod(k, M) \neq 0 \\ M & \text{otherwise} \end{cases}$
6:    $D^c_{k^*j} \leftarrow$ Use Eq. (4.7) to estimate the average end-to-end packet delay in the command path
7:    $D^f_{jk^*} \leftarrow$ Use Eq. (4.8) to estimate the average end-to-end packet delay in the feedback path
8:    **if** $\max\{D^c_{k^*j}, D^f_{jk^*}\} \leq D_0$ **then**
9:      $z_{jk^*} \leftarrow 1$
10:      $S_{k^*} \leftarrow S_{k^*} \cup \{J_j\}$
11:    **else**
12:      $k^* \leftarrow \underset{O_k \in \mathcal{M}}{\arg\min} \left\{ \max\{D^c_{kj}, D^f_{jk}\} | k = 1, 2, \ldots, M \right\}$
13:      $z_{jk^*} \leftarrow 1$
14:      $S_{k^*} \leftarrow S_{k^*} \cup \{J_j\}$
15:    **end if**
16: **end for**
17: **for** $k = 1$ to M **do**
18:    $t \leftarrow 0$
19:    **while** $S_k \neq \emptyset$ **do**
20:      $J_{j^*} = \underset{J_j \in S_k}{\arg\min} \left\{ \frac{D_j}{\Omega_j} \right\}$
21:      $S_{j^*} \leftarrow t$
22:      $C_{j^*} \leftarrow S_{j^*} + T_{j^*}$
23:      $t \leftarrow C_{j^*}$
24:      $S_k \leftarrow S_k \setminus \{J_{j^*}\}$
25:    **end while**
26: **end for**
27: **return** $S_j, C_j, z_{jk}, \forall j = 1, \ldots, N, k = 1, \ldots, M$

Source: Ebrahimzadeh and Maier (2019) © 2019 IEEE.

command and feedback paths (see lines 7–14 in Algorithm 5). This solves the assignment sub-problem. Next, the proposed CAPSTA algorithm tackles the scheduling subproblem to HOs. Toward this end, among unscheduled tasks, we first select the task with the minimum amount of $\frac{D_j}{\Omega_j}$ and then schedule it when the HO becomes available (see lines 16–25 in Algorithm 5). This, as a result, gives preference to the tasks with larger weights and shorter due times.

4.4.3 Complexity Analysis

In the proposed CAPSTA algorithm, partitioning the given task set $\mathcal{J}$ into M subsets returns a solution with complexity $\mathcal{O}(N \log N) + \mathcal{O}(N) = \mathcal{O}(N \log N)$. Next, CAPSTA solves the scheduling subproblem with time complexity $\mathcal{O}(\left\lceil \frac{N}{M} \right\rceil \log \left\lceil \frac{N}{M} \right\rceil) +$ $\mathcal{O}(N \cdot M)$. The overall time complexity is thus calculated as $\mathcal{O}(N \log N) + \mathcal{O}(M \cdot N)$, which reduces to $\mathcal{O}(N \log N) + \mathcal{O}(N^2)$ since $M \ll N$.

4.5 Delay Analysis

Recall from above that in order to ensure the quality-of-control of local/nonlocal teleoperation loops, the average end-to-end delay of HO–TOR pairs should not exceed a given threshold. Thus, in order to ensure the proper performance of our proposed CAPSTA algorithm, it is of vital importance to estimate the connection delay between any given TOR and the available HOs in both command and feedback paths. Toward this end, we develop our analytical framework to estimate the average end-to-end packet delay of local and nonlocal teleoperation in FiWi-based Tactile Internet infrastructures. In our analysis, we make the following assumptions:

- *Single-hop* wireless local area network (WLAN): MUs, HOs, and TORs are directly associated with an (optical network unit) ONU-AP via a wireless single hop, whereby ONU-MPPs serve as ONU-APs.
- *Haptic traffic model*: In both command and feedback paths, HOs and TORs transmit their update packets at a rate of 1000 packets/s with fixed deterministic interarrival times set to 1 ms (Steinbach et al., 2012).
- *Background traffic model*: MUs generate background Poisson traffic with mean packet rate λ_B (in packets/second). In addition, the background traffic rate generated by ONUs with attached fixed (wired) subscribers that are directly connected to the backhaul EPON is set to $\lambda_{ONU} = \alpha_{PON} \lambda_B$, where α_{PON} is a traffic scale factor.

For notational convenience, let us use the term "WiFi user" for all MUs, HOs, and TORs within the coverage area of an ONU-AP. We model each WiFi user as a GI/G/1 queue to account for the different packet interarrival time distributions under consideration (i.e. Poisson for background traffic and deterministic for haptic traffic). While the GI/G/1 queuing model requires the fewest assumptions among other models, it yields quite conservative results in that we can obtain only an upper bound for the average delay experienced by any packet. An accurate analysis of GI/G/1 queues can be done by solving the Lindley's integral equation in Lindley (1952). Closed-form solutions, however, are difficult to obtain, except for some known distributions. Therefore, we use the approximation method presented in Buzacott (1996) to estimate the upper bound of the average packet delay.

Let the delay experienced by any packet generated by a WiFi user be denoted by random variable D, which is the sum of the queuing delay D_Q and service time (channel access delay) D_S. To begin with, let the number of packets in the system (i.e. queue and server) be denoted by N_t, which is approximated as

$$\mathbb{E}(N_t) \approx \left(\frac{\rho^2 \left(1 + C_s^2 \right)}{1 + \rho^2 C_s^2} \right) \left(\frac{C_a^2 + \rho^2 C_s^2}{2(1 - \rho)} \right) + \rho \qquad (4.9)$$

where C_s and C_a denote the coefficient of variation of service and interarrival times, and ρ denotes the server utilization. According to Little's law, the average delay experienced by an arbitrary packet since the time it arrives in the queue until it successfully departs service is then calculated as

$$\mathbb{E}(D) = \frac{\mathbb{E}(N_t)}{\lambda} = \overbrace{\frac{1}{\lambda} \left(\frac{\rho^2 \left(1 + C_s^2 \right)}{1 + \rho^2 C_s^2} \right) \left(\frac{C_a^2 + \rho^2 C_s^2}{2(1 - \rho)} \right)}^{\text{average queuing delay } \mathbb{E}(D_Q)} + \mathbb{E}(D_S). \qquad (4.10)$$

Clearly, in order to obtain $\mathbb{E}(D)$, we need to calculate the mean service time and coefficient of variation of service time. This requires to obtain the first and second moments of service time D_S.

To compute the first and second moments of channel access delay D_S in Eq. (4.10), we defined the two-dimensional Markov chain $(s(t), b(t))$ shown in Figure 2.12 under unsaturated traffic conditions and estimated the average service time $\mathbb{E}(D_S)$ and service time variance $\mathbb{VAR}(D_S)$ in a WLAN using IEEE 802.11 distributed coordination function (DCF) for access control (see Chapter 2 for further details). We then obtain the first and second moments of the packet delay as follows:

$$\mathbb{E}(D_S) = \sum_{k=0}^{\infty} p_e^k \cdot \left(1 - p_e \right) \cdot \left[\sum_{j=0}^{\infty} p_c^j \left(1 - p_c \right) \right.$$

$$\left. \left(\left(\sum_{b=0}^{k+j} \frac{2^{\min(b,m)} W_0 - 1}{2} E_s \right) + jT_c + kT_e + T_s \right) \right] \qquad (4.11)$$

$$\mathbb{VAR}(D_S) = \sum_{k=0}^{\infty} p_e^k \left(1 - p_e \right) \sum_{j=0}^{\infty} p_c^j \left(1 - p_c \right) Q^2(j, k) - \mathbb{E}^2(D_S)$$

with

$$Q(j, k) = \left(\sum_{b=0}^{k+j} \frac{2^{\min(b,m)} W_0 - 1}{2} E_s \right) + jT_c + kT_e + T_s. \qquad (4.12)$$

After finding $\mathbb{E}(D_S)$ and $\mathbb{VAR}(D_S)$ and given the incoming traffic rate λ, we can now compute the average delay experienced by any packet for a given WiFi subscriber using Eq. (4.10).

Next, we calculate the average delay experienced by an arriving packet at the EPON backhaul. In doing so, we build on the analytical frameworks presented in Beyranvand et al. (2017) and Aurzada et al. (2014). We first define the backhaul downstream traffic intensity ρ^u and ρ^d for a time division multiplexing (TDM) passive optical network (PON) ($\Lambda = 1$) and a wavelength division multiplexing (WDM) PON ($\Lambda > 1$) as

$$\rho^u = \frac{\overline{L}}{\Lambda \cdot c_{\text{PON}}} \sum_{q=1}^{O} \sum_{i=0}^{O} \Gamma_{qi}^{\text{PON}} < 1 \tag{4.13a}$$

$$\rho^d = \frac{\overline{L}}{\Lambda \cdot c_{\text{PON}}} \sum_{q=0}^{O} \sum_{i=1}^{O} \Gamma_{qi}^{\text{PON}} < 1 \tag{4.13b}$$

where c_{PON} denotes the PON data rate, O denotes the number of ONUs, and Γ_{qi}^{PON} represents the traffic rate (in packets/second) between PON nodes q and i (with $q = 0$ denoting the OLT).

Similar to Aurzada et al. (2014), the upstream delay, D_{PON}^u, and downstream delay, D_{PON}^d, of both TDM and WDM PONs are obtained as

$$D_{\text{PON}}^u = \Phi(\rho^u, \overline{L}, \varsigma^2, c_{\text{PON}}) + \frac{\overline{L}}{c_{\text{PON}}} + 2\tau_{\text{PON}} \frac{2 - \rho^u}{1 - \rho^u} - B^u \tag{4.14}$$

$$D_{\text{PON}}^d = \Phi(\rho^u, \overline{L}, \varsigma^2, c_{\text{PON}}) + \frac{\overline{L}}{c_{\text{PON}}} + \tau_{\text{PON}} - B^u \tag{4.15}$$

where τ_{PON} denotes the average propagation delay between ONUs and OLT, $\Phi(\cdot)$ is the average queuing delay of an M/G/1 queue characterized by the Pollaczek–Khintchine formula as

$$\Phi(\rho, \overline{L}, \varsigma^2, c) = \frac{\rho}{2c(1 - \rho)} \left(\frac{\varsigma^2}{\overline{L}} + \overline{L} \right) \tag{4.16}$$

and

$$B^d = B^u = \Phi\left(\frac{\overline{L}}{\Lambda \cdot c_{\text{PON}}} \sum_{q=1}^{O} \sum_{i=1}^{O} \Gamma_{qi}^{\text{PON}}, \overline{L}, \varsigma^2, c_{\text{PON}} \right). \tag{4.17}$$

In the following, we proceed to estimate the elements of the command delay matrix $\mathbf{D}_c$ and feedback delay matrix $\mathbf{D}_f$, accounting for both local and nonlocal teleoperation scenarios.

4.5.1 Local Teleoperation

If HO O_k and the TOR that is collocated with task J_j are both associated with the same ONU-AP, the average end-to-end packet delay D_{kj}^c, $\forall k = 1, 2, \dots, M$ and

$j = 1, 2, \ldots, N$, in the command path is estimated as

$$D^c_{kj} = \mathbb{E}(D_{O_k}) + \mathbb{E}(D_{\text{ONU-AP}_n}) \tag{4.18}$$

where $\mathbb{E}(D_X)$ for a given WiFi subscriber X is obtained from Eq. (4.10) and ONU-AP$_n$ denotes the ONU-AP, which HO k and TOR j are connected to.

The average end-to-end packet delay D^f_{jk}, $\forall j = 1, 2, \ldots, N$ and $k = 1, 2, \ldots, M$, in the feedback path is then estimated as

$$D^f_{jk} = \mathbb{E}(D_{\text{TOR}_j}) + \mathbb{E}(D_{\text{ONU-AP}_n}). \tag{4.19}$$

Note that in local teleoperation, the average end-to-end delay in command and feedback paths may, in general, be different due to different traffic patterns/rates, bit error probabilities, and medium access control (MAC) settings, among others.

4.5.2 Nonlocal Teleoperation

Unlike local teleoperation, nonlocal teleoperation is carried out, if HO O_k and the TOR that is collocated with task J_j are associated with different ONU-APs. The average end-to-end packet delay D^c_{kj}, $\forall k = 1, 2, \ldots, M$ and $j = 1, 2, \ldots, N$, in the command path is therefore estimated as

$$D^c_{kj} = \mathbb{E}(D_{O_k}) + D^u_{\text{PON}} + D^d_{\text{PON}} + \mathbb{E}(D_{\text{ONU-AP}_{n'}}) \tag{4.20}$$

which accounts for the average upstream delay D^u_{PON} and downstream delay D^u_{PON} in the backhaul EPON given in Eqs. (4.14) and (4.15), respectively. Also note that ONU-AP$_{n'}$ denotes the ONU-AP with which HO O_k is associated.

The average end-to-end packet delay D^f_{jk}, $\forall j = 1, 2, \ldots, N$ and $k = 1, 2, \ldots, M$, in the feedback path is then estimated as

$$D^f_{jk} = \mathbb{E}(D_{\text{TOR}_j}) + D^u_{\text{PON}} + D^d_{\text{PON}} + \mathbb{E}(D_{\text{ONU-AP}_{n'}}). \tag{4.21}$$

4.6 Results

In this section, we examine our proposed CAPSTA algorithm. In our simulations, the task operation time T_j is sampled from a discrete uniform distribution over the range of $[10, 30]$ seconds. The delay threshold D_0 is set to 10 ms. The weight Ω_j is randomly chosen from $\{1, 2, 3, 4\}$ (i.e. four different classes). The due times are randomly chosen from $\alpha \cdot [1, \lceil \frac{1}{M} \sum_{j=1}^{N} T_j \rceil]$, where $\lceil x \rceil$ denotes the smallest integer greater than or equal to x. Each point shown in the following results is averaged over 50 randomly generated problem instants and falls within the 95% confidence interval. We compare the performance of our proposed CAPSTA algorithm with a benchmark *random assignment and scheduling* (RAS) algorithm, where for a given task an HO is randomly selected from the pool of available ones (Brucker, 2007). As for the underlying FiWi-enhanced mobile network architecture, we apply the same default parameter settings of IEEE 802.11n DCF as listed in Table I in Beyranvand et al. (2017). We consider four ONU-APs and four conventional ONUs, each serving fixed (wired) subscribers that are all involved in nonlocal H2H communications among each other. MUs and fixed subscribers generate background traffic at a mean rate of λ_B and $\alpha_{PON} \cdot \lambda_B$, respectively, whereas HOs and TORs generate haptic traffic at a fixed rate of 1000 packets/s. We consider 6-DoF (degrees-of-freedom) TORs.

First, we present the average weighted completion time (AWCT) and maximum tardiness vs. total number of available HOs M in Figures 4.3 and 4.4, respectively. We observe from Figure 4.3 that increasing M results in an exponential decrease of AWCT in both RAS and proposed CAPSTA algorithms. Specifically, in the proposed CAPSTA algorithm, increasing M from 1 to 3 results in a 67% reduction, whereas increasing M from 3 to 5 results in only a 41% reduction of AWCT. Further, we note that the proposed CAPSTA algorithm achieves a 15–27% reduction of AWTC compared to the RAS algorithm. Although achieving a lower AWCT, the beneficial impact of the proposed CAPSTA algorithm compared to the RAS algorithm is more pronounced in terms of the maximum tardiness, as shown in Figure 4.4. We observe that the proposed CAPSTA algorithm achieves

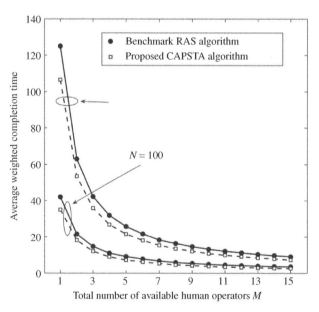

Figure 4.3 Average weighted completion time of tasks vs. total number of available human operators M ($\alpha = 1$ fixed). Source: Ebrahimzadeh and Maier (2019). © 2019 IEEE.

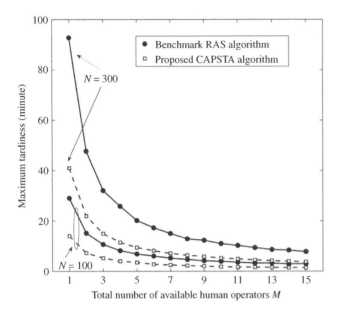

Figure 4.4 Maximum tardiness of tasks vs. total number of available human operators M ($\alpha = 1$ fixed). Source: Ebrahimzadeh and Maier (2019). © 2019 IEEE.

a 49–56% reduction of maximum tardiness. Specifically, for $N = 300$, in order to keep the maximum tardiness below 25 minutes, a total number of five HOs is needed in the RAS algorithm, whereas in the proposed CAPSTA algorithm, only two HOs are sufficient to achieve the same performance level. Further, if the decision-maker likes to keep the maximum tardiness below 10 minutes, then the number of required HOs is 5 and 12 in the proposed CAPSTA and RAS algorithms, respectively, thus achieving a notable saving in OPEX, to be examined shortly.

Next, we investigate the impact of increasing due time on the portion of the total tasks of different classes that are subject to tardiness. Toward this end, let us define R_Ω as the rate of tardy tasks with weight Ω to the total number of tasks in the same class. We have considered four different priority classes **A**, **B**, **C**, and **D**, which are associated with weight W equal to 1, 2, 3, and 4, respectively. The results of R_Ω vs. α are shown in Figure 4.5. First, for small average task due times, 94% of class **D** tasks cannot be accomplished within the expected due times, thus most of the tasks are regarded as tardy tasks. This figure is >98% for class **A**–**C** tasks. Nevertheless, as the average given due time increases, the portion of class **D** tasks that are subject to tardiness decreases exponentially. Specifically, for $\alpha = 1$, R_Ω drops below 2%. Second, we find that the proposed CAPSTA algorithm schedules the tasks in favor of high-priority ones, especially for α greater than 0.5, as shown in Figure 4.5. We

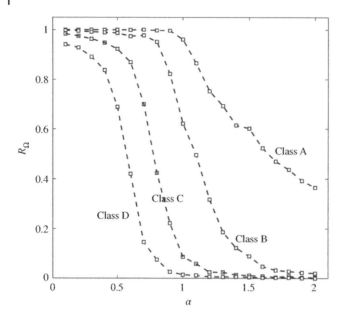

Figure 4.5 Rate R_Ω of tardy tasks vs. α for different task classes ($N = 300$ and $M = 5$ fixed). Source: Ebrahimzadeh and Maier (2019). © 2019 IEEE.

note that for α equal to 2, R_Ω converges to $< 2\%$ for classes **B–D**, whereas 36% of class **A** tasks (i.e. low-priority tasks) are still subject to tardiness.

Figure 4.6 depicts the average OPEX per task vs. total number of available HOs M for both the proposed CAPSTA and benchmark RAS algorithms. Overall, the proposed CAPSTA algorithm outperforms the RAS algorithm in terms of average OPEX per task, especially when the number of tasks is large, i.e. $N = 300$. For $M = 1$, comparing the performance of the proposed CAPSTA algorithm with the benchmark RAS algorithm, we observe a 75.3% and 78.9% reduction of average OPEX per task for $N = 100$ and $N = 300$, respectively. As M increases, the OPEX savings of the CAPSTA algorithm with respect to RAS algorithm decreases until both curves converge. The reason for this is that when the total number of available HOs M is small, the incurred OPEX is mainly due to tardy tasks, which are penalized proportional to the weighted amount of tardiness. Figure 4.6 demonstrates that the efficient scheduling of the proposed CAPSTA algorithm reduces the number of high-priority tasks that are subject to tardiness, thus achieving a significant reduction of the average OPEX per task, compared to that of the benchmark RAS algorithm.

More importantly, Figure 4.6 gives us further insights into selecting an optimal number of HOs, which should be, on the one hand, large enough to reduce the number of high-priority tardy tasks, and, on the other hand, small enough to avoid

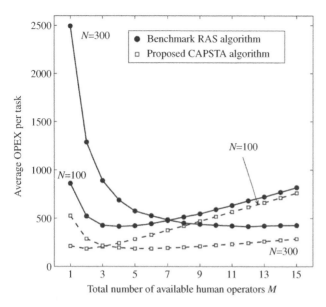

Figure 4.6 Average operational expenditures (OPEX) per task vs. total number of available human operators M ($\epsilon_h = 1$, $\epsilon_m = 5000$, and $\alpha = 1$ fixed). Source: Ebrahimzadeh and Maier (2019). © 2019 IEEE.

incurring excessive OPEX due to activating new teleoperation sessions. For the proposed CAPSTA algorithm, the optimal number of available HOs $M^\star$ that minimizes $C(X)$ is 2 and 5 for $N = 100$ and $N = 300$, respectively. We note that for the proposed CAPSTA algorithm with $M < 3$, the average OPEX per task for $N = 100$ is less compared to that of $N = 300$. Both curves meet at $M = 4$ and then the OPEX per task for $N = 100$ grows larger than that of $N = 300$. The reason for this is that for $N = 100$, while increasing M does not result in a further decrease of tardiness, it does result in an excessive increase of OPEX due to the incurred activation costs of new teleoperation sessions. In contrast, for $N = 300$, a large portion of the tasks are subject to tardiness, thus increasing M reduces the incurred OPEX due to tardiness, which in turn partly compensates for the incurred OPEX due to activating teleoperation sessions.

The average OPEX per task vs. M for different $\alpha \in \{0.1, 0.5, 1, 2\}$ for a fixed $N = 100$ is illustrated in Figure 4.7, where we examine the impact of increasing average task due times on the OPEX performance of our proposed CAPSTA algorithm. We find that for $\alpha = 0.1, 0.5$, and 1, the average OPEX, $C(X)$, is a convex function of M, having a minimum at $M^\star = 6, 4$, and 2, respectively, compared to that of $\alpha = 2$, where $C(X)$ increases linearly for increasing M, as explained above. We note that for relatively relaxed due times (i.e. $\alpha = 2$), the contribution of the

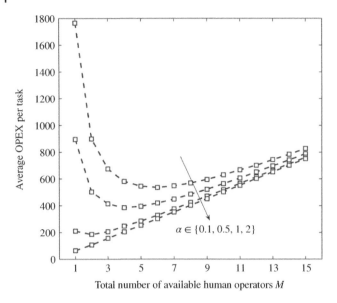

Figure 4.7 Average operational expenditures (OPEX) per task vs. total number of available human operators M ($\epsilon_h = 1$, $\epsilon_m = 5000$, and $N = 100$ fixed). Source: Ebrahimzadeh and Maier (2019). © 2019 IEEE.

incurred penalty due to task tardiness is negligible compared to that of activating excessive teleoperation sessions. Hence, the average OPEX grows linearly as M increases. Therefore, when the average due time is large, it is beneficial to perform the teleoperation tasks by only one HO, provided that proper scheduling is fulfilled (see Figure 4.7).

Next, the average OPEX vs. operational cost, ϵ_h, per time unit of tardiness is shown in Figure 4.8, which renders the following interesting insights. First, Figure 4.8 specifies the range of ϵ_h for which the proposed CAPSTA algorithm achieves more beneficial results in terms of OPEX for two given $M = M_1$ and $M = M_2$. For instance, while $M = 2$ always leads to a smaller OPEX per task compared with that of $M = 1$, decreasing the number of HOs from $M = 5$ to $M = 2$ does not achieve such reduction. To be more specific, $M = 5$ is more OPEX-beneficial than $M = 2$ only if ϵ_h is greater than 0.5. This, however, is a quite counterintuitive observation whether or not increasing M results in OPEX savings depends not only on the average task due times (as explained before) but also the operational cost, ϵ_h, per time unit of tardiness, as increasing M from 2 to 10 only incurs an additional OPEX due to activating new teleoperation sessions. Further, we also note that for $M = 10$, the rate at which $C(\mathbf{X})$ increases degrades as ϵ_h grows. This is due to the fact that for a large M (e.g. $M = 10$) OPEX is less likely due to task tardiness, thus increasing ϵ_h does not increase $C(\mathbf{X})$ significantly,

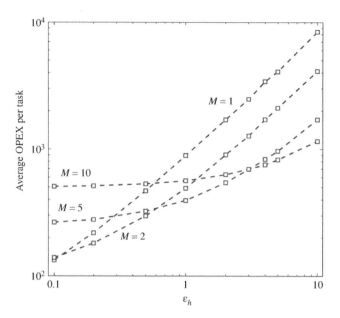

Figure 4.8 Average operational expenditures (OPEX) per task vs. ϵ_h for different number of available human operators $M \in \{1, 2, 5, 10\}$ ($\alpha = 0.5$, $\epsilon_m = 5000$, and $N = 100$ fixed). Source: Ebrahimzadeh and Maier (2019). © 2019 IEEE.

as opposed to small values of M (e.g. $M = 1$), where increasing ϵ_h results in a significant increase of $C(\mathbf{X})$. As a result, Figure 4.8 together with Figure 4.7 are instrumental in helping optimize OPEX for a given set of system parameter values.

Next, we examine the impact of average task due times on the OPEX performance of the proposed CAPSTA algorithm. Figure 4.9 presents the average OPEX per task vs. α, which reflects the amount of average task due time. For $M = 1$, 2, and 5, the average OPEX per task decreases for increasing α and levels off for $\alpha > 1$. This is due to the fact that for smaller values of M, OPEX is mainly due to penalizing the tardy tasks. Therefore, increasing α translates into a reduced average tardiness, thus alleviating the average OPEX per task. On the other hand, for large values of M (e.g. $M = 10$), the contribution of the first term of Eq. (4.6) to OPEX is greater than the second term. For this reason, we do not observe a notable decrease of the OPEX as α increases.

Finally, we evaluate the end-to-end delay performance of local and nonlocal teleoperation. Figure 4.10 depicts the average end-to-end packet delay of local teleoperation vs. mean background traffic λ_B for different $N_{\mathrm{MU}} \in \{2, 3, 4, 5, 10\}$, where N_{MU} denotes the number of MUs that reside within the coverage of each ONU-AP.

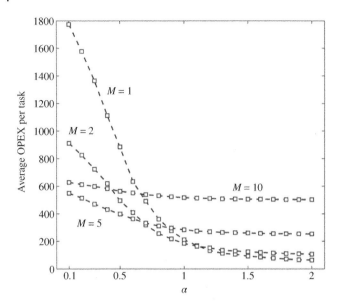

Figure 4.9 Average operational expenditures (OPEX) per task vs. α for different number of available human operators $M \in \{1, 2, 5, 10\}$ ($\epsilon_h = 1$, $\epsilon_m = 5000$, and $N = 100$ fixed). Source: Ebrahimzadeh and Maier (2019). © 2019 IEEE.

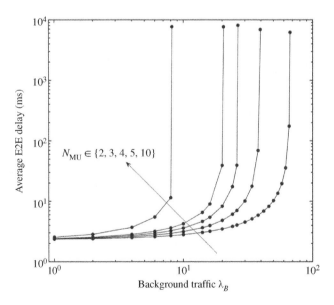

Figure 4.10 Average end-to-end packet delay of local teleoperation vs. background traffic rate λ_B for different $N_{MU} \in \{2, 3, 4, 5, 10\}$. Source: Ebrahimzadeh and Maier (2019). © 2019 IEEE.

We find that an average end-to-end delay of 2.5 ms is achievable for local teleoperation involved HO–TOR pairs. It is worthwhile to mention that the amount of time waited by packets in the second hop (i.e. MAC queue of the ONU-AP) is notably larger than that of the first hop (i.e. MAC queue of the HO/TOR), which is a direct consequence of the high incoming packet rate at the ONU-AP. Second, we observe that for $N_{MU} \in \{2, 3, 4, 5\}$, the average end-to-end delay remains under 10 ms for a wide range of background traffic rate λ_B.

Figure 4.11 illustrates the average end-to-end packet delay of nonlocal teleoperation vs. background traffic rate λ_B for different values of $\alpha_{PON} \in \{100, 200, 500\}$ and $N_{MU} = 2$. In nonlocal teleoperation, the obtained end-to-end delay is as low as 2.8 and 4.5 ms for $l_{PON} = 20$ and 100 km (compared to 2.5 ms in local teleoperation). Further, we observe that for a given background traffic load, say $\lambda_B = 10$ packets/s, increasing l_{PON} from 20 to 100 km results in a 1.6 ms increase of the end-to-end delay from 3.5 to 5.1 ms. This is counterintuitive in that the 80 km increase of backhaul fiber length accounts for only 267 μs in propagation delay, which is much smaller than 1.6 ms. The reason for this lies in the impact of increasing l_{PON} on the delay performance of the multipoint control protocol (MPCP) used in the backhaul EPON.

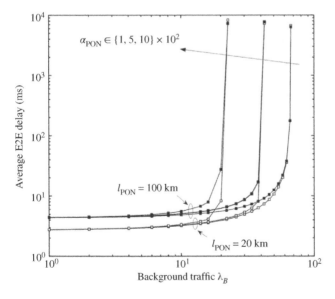

Figure 4.11 Average end-to-end packet delay of nonlocal teleoperation vs. background traffic rate λ_B for different $\alpha_{PON} \in \{100, 200, 500\}$ ($N_{MU} = 2$ fixed). Source: Ebrahimzadeh and Maier (2019). © 2019 IEEE.

4.7 Discussion

As the Tactile Internet emerges, the flows generated by different applications become more diverse, each requiring a different (quality of service) QoS/E. To overcome the issues arising from traditional network management models, including limited reconfigurability and complex per-flow traffic management, software-defined networking (SDN)/network function virtualization (NFV) is a promising solution, where a clear distinction is made between the control and data planes. This as a result can provide the task coordinator with a logically centralized overview of the whole network, gather application-dependent requirements (teleoperation in our studied scenario), and reconfigure network parameters to achieve the desired QoS/E. In this context, NFV is a promising technique, which can be used not only to further reduce the capital expenditure (CAPEX) and OPEX issues of teleoperation over FiWi-enhanced mobile networks but also to support a wider variety of HSI and TOR types (see Figure 2.2). More importantly, given that FiWi-enhanced mobile networks have to cope with the seamless integration of both optical and wireless subnetworks, the role of SDN is even more pronounced in alleviating the difficulties of network design, control, and management, especially with the coexistence of different types of traffic (Liu et al., 2016). In this context, Thyagaturu et al. (2016) presents a thorough review of the studies that examine the SDN paradigm in optical networks, also referred to as software-defined optical networks (SDONs). While the concept of sotfwarization of network protocols realized via SDN enables the study of new ideas and optimization models, thereby significantly reducing the deployment costs and speeding up the upgrade process, virtualization facilitates service migration, thus allowing for location-aware service provisioning in a cost-efficient manner (Cabrera et al., 2019).

4.8 Conclusion

We investigated the performance of our proposed CAPSTA algorithm in solving the prioritized assignment and scheduling of delay-constrained teleoperation tasks in FiWi-enhanced Tactile Internet network infrastructures. The obtained results show that the proposed algorithm reduces the average weighted task completion time, maximum tardiness, and average OPEX, compared to the benchmark RAS algorithm. Specifically, the proposed CAPSTA algorithm achieves a 15–27% reduction of average weighted task completion time and a 49–56% reduction of maximum tardiness. In addition, compared to the benchmark RAS algorithm, the proposed CAPSTA algorithm achieves a 75.3% and 78.9% reduction of average OPEX per task for $N = 100$ and $N = 300$, respectively. Our results also give insights into finding the optimal number of HOs to minimize the average OPEX per completed task for different deployment scenarios. More precisely, we have shown that for the proposed CAPSTA algorithm, the optimal number of available HOs $M^\star$ that minimizes OPEX is 2 and 5 for $N = 100$ and $N = 300$, respectively. Finally, we have shown that the considered solution is able to achieve an average end-to-end packet delay of < 10 ms for both local and nonlocal teleoperation for a wide range of background traffic rates. An interesting future research avenue is to investigate the role of virtualization in FiWi-enhanced mobile networks to eliminate the physical layer interaction of the often heterogeneous Tactile Internet applications, thus realizing an infrastructure/technology-independent architecture.

5

Cooperative Computation Offloading in FiWi-Enhanced Mobile Networks

5.1 Introduction

To address the contradiction between the rapid increase of computationally intensive, delay-sensitive applications (e.g. Tactile Internet, augmented reality (AR)/virtual reality (VR), and interactive gaming) and resource-limited smart mobile devices, mobile cloud computing (MCC) has emerged to reduce the computational burden of mobile devices and broaden their capabilities by extending the concept of cloud computing to the mobile environment via full and/or partial computation offloading. Even though MCC allows mobile devices to benefit from powerful computing resources to save battery power and accelerating task execution, it raises several technical challenges due to additional communication overhead and poor reliability that remote computation offloading may introduce. To overcome these limitations, *multi-access edge computing (MEC)* has recently emerged to provide cloud computing capabilities at the edge of access networks, leveraging the physical proximity of edge servers and mobile users (MUs) to achieve reduced communication latency and increased reliability (Chen et al., 2016).

While a conventional (remote) cloud provides high storage and computational capabilities, it may pose large latency due to communications, as it is usually physically distant from the MUs. On the other hand, MEC may offer a reduced communication-induced latency, but it may pose an excessive processing latency due to limited computational capabilities. In a broader vision, remote cloud and MEC servers can thus coexist and be complementary to each other, giving rise to *cooperative computation offloading*. The ultimate goal of MEC, in fact, is to achieve an ultra-low response time, which is defined as the time interval between the time instant at which a task is released from a mobile device until it is processed (either locally or remotely) and the result is received by the device. This time interval may include the waiting (queueing) and processing times in

Toward 6G: A New Era of Convergence, First Edition. Amin Ebrahimzadeh and Martin Maier.
© 2021 The Institute of Electrical and Electronics Engineers, Inc.
Published 2021 by John Wiley & Sons, Inc.

either the local central processing unit (CPU) or edge/remote server as well as the communication latency between the mobile device and edge/remote cloud. Given the additional communication overhead that offloading introduces, a key technical challenge is to find a trade-off between the cost of computation and communication to enhance user experience in terms of lower latency and energy consumption. In this chapter, motivated by Xiao and Krunz (2018), we focus on the quality of experience (QoE) of MUs measured by the average response time that can be influenced by the queueing/processing and transmission delay components, including those between MUs and MEC servers and also between MEC servers and the remote cloud.

To achieve the desired energy-delay performance, the so-called dynamic voltage scaling (DVS) is a promising technique that varies the supply voltage and clock frequency based on the computation load to achieve a suitable trade-off between task execution time and energy consumption (Wang et al., 2016). While computation offloading mainly relies on the computational capabilities of the edge/remote servers, the DVS technique enables the MUs to adaptively adjust their computational speed to reduce energy consumption or shorten task execution time. Therefore, incorporating the DVS technique into computation offloading offers more flexibility at the device side, enabling MUs to achieve self-awareness via a design approach commonly known as *self-organization* to further improve their QoE under different scenarios (Klaine et al., 2017).

5G mobile networks have led to an increasing integration of cellular and WiFi technologies and standards, giving rise to so-called HetNets, which mandates the need for addressing the backhaul bottleneck challenge (Maier and Ebrahimzadeh, 2019). Recently, we have explored the performance gains obtained from unifying coverage-centric 4G long-term evolution-advanced (LTE-A) HetNets and capacity-centric FiWi access networks based on data-centric Ethernet technologies with resulting fiber backhaul sharing and WiFi offloading capabilities toward realizing 5G networks (Beyranvand et al., 2017). By means of probabilistic analysis and verifying simulations based on recent and comprehensive smartphone traces, we showed that an average end-to-end latency of < 10 ms can be achieved for a wide range of traffic loads and that MUs can be provided with highly fault-tolerant FiWi connectivity for reliable low-latency fiber backhaul sharing and WiFi offloading. Note, however, that only data offloading was considered in Beyranvand et al. (2017) without any computation offloading via MEC. Furthermore, the feasibility of implementing conventional cloud and MEC in FiWi access networks was investigated in Rimal et al. (2017c), where the main objective was to design a unified resource management scheme to integrate offloading activities with the underlying FiWi operations. While much of the effort in these papers has been devoted to the management of networking resources, cooperation between mobile devices, MEC servers, the remote cloud and the problem of offloading decision making have not been investigated. In Tan et al. (2017), a scalable online algorithm for task scheduling in an edge-cloud system was proposed, which was verified by simulations using real-world traces from Google. A hierarchical MEC-based architecture was presented in Tong et al. (2016) with a focus on the workload placement problem. In Chen et al. (2017), an optimization framework was presented for solving the problem of joint offloading decision and allocation of computation and communication resources with the aim of minimizing a weighted sum of the costs of energy, cost of computation, and the delay for all users. More recently, Xiao and Krunz (2018) studied the computation offloading problem for cooperative fog computing networks and investigated the fundamental trade-off between QoE of MUs and power efficiency of fog nodes. In Guo and Liu (2018), a collaborative computation offloading scheme for MEC over FiWi networks was presented. All mentioned papers, however, mainly focused on the management of computing resources without further investigating the impact of the capacity-limited backhaul.

In this chapter, we examine the performance gains obtained by cooperative computation offloading in MEC-enabled FiWi enhanced HetNets, which relies on not only the computational capabilities of edge/cloud servers but also the limited local computing resources at the device side. More specifically, we aim to design a two-tier MEC-enabled FiWi enhanced HetNet architecture, where the mobile devices as well as the edge servers cooperatively offload their computation tasks toward achieving a reduced average response time. We take into account both crucial aspects of limitations stemming from communications and computation in our design approach via accurate modeling of the fronthaul/backhaul as well as edge/cloud servers, while paying particular attention to offloading decision-making between MUs and edge servers as well as edge servers and the remote cloud. Another important aspect of MEC is to cope with the additional complexity that may arise in such a scenario by relying, fully or partially, on the limited local computing resources of MUs when they are most needed. The inherent time-varying nature of FiWi-enhanced HetNets, which is a direct consequence of user mobility, entails exploiting a function that continuously tune the local computational capabilities of mobile devices in order to ensure an improved QoE. This can be achieved via adaptive reconfiguration of an MU given its goals, capabilities, and constraints via a design approach commonly known as self-awareness. Contributing to this effort, we leverage on the self-awareness of MUs by applying the DVS technique for making appropriate energy-delay trade-offs subject to given energy and delay constraints. In particular, the contributions of this chapter are as follows:

- We design a two-tier hierarchical MEC-enabled FiWi enhanced HetNet-based architecture for computation offloading, which leverages both local and nonlocal computing resources to achieve low response time and energy consumption for MUs. We also propose a simple but efficient offloading orchestration mechanism to achieve an improved QoE for MUs.

- We develop an analytical framework to examine the performance of a system model of our FiWi-based cooperative offloading scheme coexistent with conventional human-to-human (H2H) traffic (i.e. voice, video, and data) in terms of average response time as well as energy consumption of MUs. In our analysis, we develop detailed models of both communication and computation, incorporating WiFi/LTE-A wireless access and capacity-limited backhaul fiber links as well as resource-limited edge/remote cloud servers.
- Given the additional complexity incurred by integrating the cooperative computation offloading strategy in a FiWi enhanced HetNet architecture, any deviation from optimal delay performance is inevitable. To cope with this and in order to allow MUs to flexibly rely on their local computing resources by means of reconfiguration, we propose a self-organization framework to allow mobile devices to adaptively tune their offloading probability as well as computational capabilities via the DVS technology. The proposed self-organizing design results in a Pareto frontier characterization of the trade-off between average task execution time and energy consumption.

The remainder of the chapter is structured as follows. In Section 5.2, we present our proposed architecture of MEC-enabled FiWi enhanced HetNets and cooperative offloading mechanism. In Section 5.3, we present our analytical framework for estimating the energy-delay performance of our proposed cooperative task offloading scheme. The proposed self-organization scheme is presented in Section 5.4. Section 5.5 presents numerical results. Finally, Section 5.6 concludes the chapter.

5.2 System Model

Figure 5.1 depicts the generic architecture of the considered FiWi enhanced LTE-A HetNets. We equip selected optical network unit-base stations (ONU-BSs)/mesh portal points (MPPs) with MEC servers (or simply called *edge servers* hereafter) collocated at the optical-wireless interface. MUs may offload fully or portion of their incoming computational tasks to nearby edge servers. In addition to edge servers, the optical line terminal (OLT) is equipped with cloud computing facilities, which consist of multiple servers dedicated to processing mobile tasks. Each MU uses a task scheduler that decides whether to offload a task to an edge server or execute it locally in its local CPU. We model the task scheduler in each MU by a queueing system, as illustrated in Figure 5.2. Each MU uses a first in, first out (FIFO) task scheduler that decides whether to offload a task to an edge server or execute it locally in its local CPU. We model the task scheduler in each MU by a FIFO queueing system, as illustrated in Figure 5.2. We assume that in each mobile

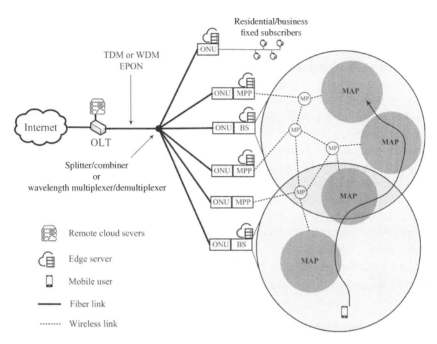

Figure 5.1 Generic multi-access edge computing (MEC)-enabled fiber-wireless (FiWi) enhanced LTE-Advanced (LTE-A) Heterogenous Networks (HetNets) architecture. Source: Ebrahimzadeh and Maier (2020). © 2020 IEEE.

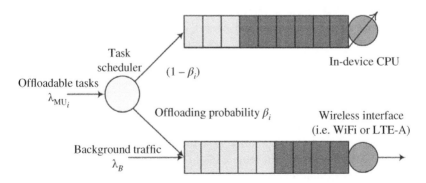

Figure 5.2 Schematic of task scheduler and queueing system for mobile user (MU) i, which includes two disjoint queues served by local central processing unit (CPU) and WiFi/LTE-A wireless interface. Source: Ebrahimzadeh and Maier (2020). © 2020 IEEE.

device there are two servers, namely, the CPU and the wireless interface (i.e. WiFi or LTE-A). The former server is used to model the local task execution at the MU's CPU, whereas the latter is responsible for offloading tasks to an edge server in proximity. We assume that MUs generate background Poisson traffic at mean packet rate λ_B (in packets/s) (see Figure 5.2). We also assume that tasks arrive at MU i's scheduler at rate λ_{MU_i}. The task scheduler at MU i makes its decision based on the value of the so-called *offloading probability*, β_i, which is defined as the probability that an incoming task is offloaded to the edge server. Tasks generated by MU i are characterized by B_i^l and D_i^l, which denote the average size of computation input data (e.g. program codes and input parameters) and average number of CPU cycles required, respectively. Computation tasks are assumed to be atomic and thus cannot be divided into subtasks. We also assume that each edge server is equipped with a FIFO task scheduler, which decides whether to execute an incoming task or further offload it to the remote cloud. We also assume that each edge server is equipped with a task scheduler, which decides whether to execute an incoming task or further offload it to the remote cloud. Similar to MUs, a task arriving at edge server j is further offloaded to the remote cloud with probability α_j or executed locally with probability $(1 - \alpha_j)$.

5.3 Energy-Delay Analysis of the Proposed Cooperative Offloading

In this section, we analyze the performance of our proposed cooperative MEC-enabled FiWi enhanced LTE-A HetNets in terms of average response time and energy consumption for task offloading coexistent with conventional H2H traffic. Many related recent studies (e.g. (Xiao and Krunz, 2018; Chen et al., 2016; Sun and Ansari, 2017; Fan and Ansari, 2018; Liu et al., 2018; Rodrigues et al., 2017, 2018)) assumed a Poisson task arrival model and an exponentially distributed number of required CPU cycles for task execution. In this chapter, we follow the same research line and build our analysis on these assumptions. Further, tasks are assumed to be computationally intensive, mutually independent, and can be executed either locally or remotely on an edge server or the remote cloud via computation offloading.[1] Each edge server has a limited computational capability and can serve a single task at a time (Xiao and Krunz, 2018; Sun and Ansari, 2017). Besides, the remote cloud comprises a limited number of high-performance computing servers, each of which can serve a single task at

---

1 We note that computation offloading may help enable the realization of a wide variety of context-aware, computation intensive applications with low response time requirements, e.g. simultaneous localization and mapping (SLAM) and/or 3D reconstruction of the surrounding environment in an AR application.

a time. While we assume that a tail-drop queue buffer management scheme is considered, the size of the data buffers, including those deployed at optical network units (ONUs) and OLT, is assumed to be sufficiently large to avoid any packet loss due to overflow (Rimal et al., 2017a).

5.3.1 Average Response Time

In the proposed cooperative offloading scheme, both computation and communication induced latencies may contribute to the resultant average response time experienced by MUs.[2] First, we estimate the latencies due to computation for both local and nonlocal computing. For a given MU i, who is involved in task offloading, assuming i.i.d. exponentially distributed task interarrival times and given the offloading probability β_i, the tasks arriving at the CPU queue for local computing follow a Poisson process with rate $(1 - \beta_i) \cdot \lambda_{MU_i}$, whereas the offloaded tasks arriving at the wireless interface queue follow a Poisson process with rate $\beta_i \cdot \lambda_{MU_i}$. This is because thinning a Poisson process with a fixed probability results in another Poisson process. Let D_i^l be the average number of required CPU cycles to execute a task arriving at MU i. The average local task execution time τ_i^l at MU i is given by

$$D(f_i) = \tau_i^l = \frac{D_i^l}{f_i} \tag{5.1}$$

where f_i is the clock frequency (in CPU cycles per second) of MU i. Assuming that the number of required CPU cycles per task follows an exponential distribution, we can model the local CPU server of MU i as an M/M/1 queue with mean arrival rate $(1 - \beta_i) \lambda_{MU_i}$ and mean task execution time τ_i^l. The average delay Δ_{MU_i} of local task execution (which includes both queueing and service times) at MU i's CPU is then given by

$$\Delta_{MUi} = \frac{1}{\mu_i^l - (1 - \beta_i) \lambda_{MUi}} \tag{5.2}$$

where μ_i^l, which is equal to $1/\tau_i^l$, is the rate at which the executed tasks depart from MU i's CPU. We note that Eq. (5.2) is valid only if $(1 - \beta_i) \lambda_{MU_i} \tau_i^l < 1$.

Let $\mathcal{R}_j$ denote the set of MUs that are served by edge server j. Further, let $\lambda_{0,j}^e$ be the mean arrival rate and $D_{0,j}^e$ denote the required number of CPU cycles of offloaded tasks from the fixed (wired) subscribers, if any, which may be directly connected to edge server j. Given the offloading probabilities $\beta_i, \forall MU_i \in \mathcal{R}_j$, the

---

2 It is worthwhile to mention that although network bandwidth fluctuations may lead to variations of the response time, our main focus in this chapter is to estimate the long-run average of the response time.

mean arrival rate λ_{MEC_j} at the task scheduler of edge server j is computed as follows:

$$\lambda_{\mathrm{MEC}j} = \lambda^e_{0,j} + \sum_{\mathrm{MU}_i \in \mathcal{R}_j} \beta_i \cdot \lambda_{\mathrm{MU}i}. \tag{5.3}$$

Let τ^e_j denote the average task execution time at edge server j. For estimating τ^e_j, we compute the average number $\overline{D}^e_j$ of CPU cycles required to execute a task at edge sever j as follows:

$$\overline{D}^e_j = \frac{\lambda^e_{0,j} D^e_{0,j} + \sum_{\mathrm{MU}_i \in \mathcal{R}_j} \beta_i \cdot \lambda_{\mathrm{MU}i} \cdot D^l_i}{\lambda^e_{0,j} + \sum_{\mathrm{MU}_i \in \mathcal{R}_j} \beta_i \cdot \lambda_{\mathrm{MU}i}} \tag{5.4}$$

which is then used to calculate τ^e_j, which is given by

$$\tau^e_j = \frac{\overline{D}^e_j}{f^e_j} \tag{5.5}$$

where f^e_j is the computational capability (in CPU cycles per second) of edge server j. Modeling edge server j as an M/M/1 queue with mean arrival rate $(1 - \alpha_j) \lambda_{\mathrm{MEC}_j}$ and mean service time τ^e_j, the average delay Δ_{MEC_j} of task execution at edge server j is calculated as follows [3]:

$$\Delta_{\mathrm{MEC}j} = \frac{1}{\mu^e_j - (1 - \alpha_j) \lambda_{\mathrm{MEC}j}} \tag{5.6}$$

whereby $\mu^e_j = 1/\tau^e_j$. Substituting Eq. (5.3) in Eq. (5.6) provides the following expression:

$$\Delta_{\mathrm{MEC}_j} = \frac{1}{\mu^e_j - (1 - \alpha_j) \cdot \left(\lambda^e_{0,j} + \sum_{\mathrm{MU}_i \in \mathcal{R}_j} \beta_i \cdot \lambda_{\mathrm{MU}_i} \right)} \tag{5.7}$$

which is valid only if

$$\tau^e_j \cdot (1 - \alpha_j) \cdot \left(\lambda^e_{0,j} + \sum_{\mathrm{MU}_i \in \mathcal{R}_j} \beta_i \cdot \lambda_{\mathrm{MU}_i} \right) < 1.$$

Next, we proceed to estimate the task execution delay at the remote cloud. Let $\mathcal{R}$ denote the set of edge servers that are connected to the remote cloud. The mean arrival rate λ_c at the remote cloud is obtained as follows:

$$\lambda_c = \lambda^c_0 + \sum_{\mathrm{MEC}_j \in \mathcal{R}} \alpha_j \cdot \lambda_{\mathrm{MEC}j}. \tag{5.8}$$

---

3 Although an edge server may comprise a number of physical and/or virtual machines to process the incoming tasks, we are focusing on a coarse-grained scenario, thus modeling an edge server as a single entity, as in many related works, e.g. Xiao and Krunz (2018), Sun and Ansari (2017), and Fan and Ansari (2018).

Let λ_0^c and D_0^c denote the arrival rate and number of CPU cycles required to execute the background tasks[4] at the remote cloud, respectively.

Further, let τ_c denote the average task execution time at the remote cloud. In order to estimate τ_c, we first calculate the average number $\overline{D}_c$ of CPU cycles required to execute a task at the remote cloud, which is given by

$$\overline{D}_c = \frac{\lambda_0^c D_0^c + \sum_{\text{MEC}_j \in R} \alpha_j \cdot \lambda_{\text{MEC}j} \cdot \overline{D}_j^c}{\lambda_0^c + \sum_{\text{MEC}_j \in R} \alpha_j \cdot \lambda_{\text{MEC}j}} \tag{5.9}$$

which is then used to estimate τ_c as follows:

$$\tau_c = \frac{\overline{D}_c}{f^c} \tag{5.10}$$

where f^c is the computational capability of each of the s homogeneous servers deployed at the remote cloud. We can thus model the remote cloud as an M/M/s queue with mean arrival rate λ_c (given by Eq. (5.8)) and mean service time τ_c (given by Eq. (5.10)). The average delay Δ_c experienced by an arbitrary task in the remote cloud is then estimated by the well-known Erlang-C formula:

$$\Delta_c = \frac{C(s, a) \cdot \tau_c}{s - a} + \tau_c \tag{5.11}$$

where the carried load a is equal to $\lambda_c \cdot \tau_c$ and $C(s, a)$ is given by

$$C(s, a) = \frac{\frac{a^s \cdot s}{s!(s-a)}}{\sum_{k=0}^{s-1} \frac{a^k}{k!} + \frac{a^s \cdot s}{s!(s-a)}} \tag{5.12}$$

which is the probability that an arriving task finds all the s servers busy. Note that Eq. (5.12) is valid only if the carried load does not exceed the number of servers (i.e. $\lambda_c \cdot \tau_c < s$).

Next, we turn our attention to calculating the communication induced latency in our cooperative task offloading scheme. Recall from above that the offloaded tasks arrive at the wireless interface of MU i with rate $\beta_i \cdot \lambda_{\text{MU}_i}$. With L_m denoting the maximum payload size of a single packet, the number of packets per task is equal to $\left\lceil \frac{B_i^l}{L_m} \right\rceil$. We can then estimate the rate Γ_{MU_i} at which packets arrive at the wireless interface of MU i as follows:

$$\Gamma_{\text{MU}_i} = \lambda_B + \left\lceil \frac{B_i^l}{L_m} \right\rceil \cdot \beta_i \cdot \lambda_{\text{MU}_i} \tag{5.13}$$

where λ_B denotes the background H2H traffic (see also Figure 5.2).

---

4 Some of the fixed subscribers may not be connected to any edge server in proximity and thus offload their tasks directly to the remote cloud via the backhaul Ethernet passive optical network (EPON). We refer to such tasks as cloud background tasks.

5.3.1.1 Delay Analysis of WiFi Users

First, we calculate the average packet delay Θ_i^{WiFi} in the uplink for MU i, who is associated with an ONU-AP through WiFi. For a given set of network model parameters, we can estimate Θ_i^{WiFi} as in Aurzada et al. (2014):

$$\Theta_i^{\text{WiFi}} = \frac{1}{\frac{1}{\Delta_i} - \Gamma_{\text{MU}_i}}; \quad \Delta_i \cdot \Gamma_{\text{MU}_i} < 1 \tag{5.14}$$

where Δ_i denotes the average channel access delay and Γ_{MU_i} is given by Eq. (5.13). Note that Eq. (5.14) accounts for both queueing and channel access (service) delay[5].

Lemma 5.1 *The average channel access delay Δ_i of MU i is obtained as follows*

$$\Delta_i = \sum_{k=0}^{\infty} p_{e,i}^k \left(1 - p_{e,i}\right)$$

$$\times \left[\sum_{j=0}^{\infty} p_{c,i}^j \left(1 - p_{c,i}\right) \left(\left(\sum_{b=0}^{k+j} \frac{2^{\min(b,m)} W_0 - 1}{2} E_s \right) + jT_{c,i} + kT_{e,i} + T_{s,i} \right) \right], \tag{5.15}$$

where $p_{e,i}$ is the probability of an erroneous transmission, $p_{c,i}$ is the probability of a collision, W_0 is the initial contention window size, E_s is the expected time-slot duration, and $T_{c,i}$, $T_{e,i}$, and $T_{s,i}$ denote the average duration of a collided, erroneous, and successful transmission of MU i, respectively.

Proof: See Appendix A.4.   □

We note that the average access delay Δ_i consists of time delays due to carrier sensing, exponential back-offs, collided and erroneous (if any) attempts, successful transmission, and acknowledgement. It is also worthwhile to mention that the presence of interfering users may increase the collision probability, $p_{c,i}$, of MU i, thus increasing its average channel access delay (see Eq. (5.15)).

5.3.1.2 Delay Analysis of 4G LTE-A Users

Next, we assume a 4G LTE-A cellular network and estimate its uplink delay. Let p_i^{tx} denote the transmission power of MU i. We use the Shannon–Hartley theorem

---

5 Similar to Beyranvand et al. (2017), Aurzada et al. (2014), Kafaie et al. (2018), Zhu et al. (2012), Liu et al. (2013), Han et al. (2006), and Pham et al. (2005), the WiFi channel access time governed by the IEEE 802.11 distributed coordination function (DCF) is assumed to be exponentially distributed. This is justified by the DCF channel access mechanism, which includes carrier sensing, binary exponential back-off(s), and reattempts (if any) due to collisions and erroneous transmissions.

to estimate the uplink data rate r_i^{LTE} of MU i transmitting to base station (BS) k via a 4G LTE-A cellular network as follows:

$$r_i^{\text{LTE}} = \omega \log_2 \left(1 + \frac{p_i^{\text{tx}} G_{i,k}}{\overline{\omega}_0^2 + \sum_{j \neq i} p_j^{\text{tx}} G_{j,k}} \right) \tag{5.16}$$

where ω and $\overline{\omega}_0^2$ are the channel bandwidth and background noise power, respectively; $G_{i,k}$ denotes the channel gain between MU i and BS k. We use (Beyranvand et al., 2017, Eq. (37)) to estimate the uplink delay of LTE-A users, which is given by

$$\Theta_i^{\text{LTE}} = \frac{\rho_{\text{BS}}^u}{2r_i^{\text{LTE}}(1 - \rho_{\text{BS}}^u)} \left(\frac{\varsigma_L^2}{\overline{L}} + \overline{L} \right) + \frac{\overline{L}}{r_i^{\text{LTE}}} + D_{\text{RA}}^{\text{up}} + D_{\text{setup}} + \tau_{\text{BS}} \tag{5.17}$$

where $D_{\text{RA}}^{\text{up}}$ is the initial random access delay (given by (Beyranvand et al., 2017, Eq. (38))), D_{setup} denotes the connection setup delay after passing the random access process successfully, ρ_{BS}^u denotes the uplink traffic intensity, τ_{BS} is the propagation delay in the cellular network, and $\overline{L}$ and ς_L^2 denote the mean and variance of the packet length, respectively. We note that, according to Eqs. (5.16) and (5.17), the achievable uplink data rate for MU i is decreased as a larger number of users is connected to the cellular BS, thereby increasing the packet delay experienced by MU i.

Each MU is directly associated with an ONU-AP or a cellular BS via a wireless single hop, whereby ONU-MPPs serve as ONU-APs. The WiFi connection and interconnection times of MUs are assumed to fit a truncated Pareto distribution, as validated via recent smartphone traces in Beyranvand et al. (2017). The probability $P_{\text{temp}}^{\text{MU}}$ that an MU is temporarily connected to an ONU-AP is estimated as $\overline{T}_{\text{on}}/(\overline{T}_{\text{on}} + \overline{T}_{\text{off}})$, whereby $\overline{T}_{\text{on}}$ and $\overline{T}_{\text{off}}$ denote the average WiFi connection and interconnection time, respectively. In this chapter, we assume that $\overline{T}_{\text{on}} = 28.1$ minutes and $\overline{T}_{\text{off}} = 10.3$ minutes, which are consistent with the measurements of PhoneLab traces (see (Beyranvand et al., 2017) for further details). With these considerations, MU i is either connected to an ONU-AP through WiFi with probability $P_{\text{temp}}^{\text{MU}}$ or an ONU-BS through cellular network with probability $(1 - P_{\text{temp}}^{\text{MU}})$. The average task transmission delay Θ_i^{UL} in the uplink is then computed as follows:

$$\Theta_i^{\text{UL}} = \left(P_{\text{temp}}^{\text{MU}} \cdot \Theta_i^{\text{WiFi}} + (1 - P_{\text{temp}}^{\text{MU}}) \cdot \Theta_i^{\text{LTE}} \right) \left\lceil \frac{B_i^l}{L_m} \right\rceil . \tag{5.18}$$

5.3.1.3 Delay Analysis of Backhaul EPON

Let D_{PON}^u denote the average packet delay in the backhaul EPON in the upstream direction. The average task transmission delay Θ^{PON} in the backhaul is then equal to $D_{\text{PON}}^u \cdot \left\lceil \frac{B_i^l}{L_m} \right\rceil$, where D_{PON}^u is given by Beyranvand et al. (2017):

$$D_{\text{PON}}^u = \Phi\left(\rho^u, \overline{L}, \varsigma_L^2, c_{\text{PON}}\right) + \frac{\overline{L}}{c_{\text{PON}}} + 2\tau_{\text{PON}} \frac{2 - \rho^u}{1 - \rho^u} - B^u \tag{5.19}$$

whereby ρ^u is the upstream traffic intensity, τ_{PON} is the propagation delay between ONUs and OLT, c_{PON} is the EPON data rate, $\Phi(\cdot)$ denotes the well-known Pollaczek–Khintchine formula, and B^u is obtained as $\Phi\left(\frac{\overline{L}}{\Lambda c_{\text{PON}}} \sum_{i=1}^{O} \sum_{q=1}^{O} \Gamma_{iq}^{\text{PON}}, \overline{L}, \varsigma_L^2, c_{\text{PON}}\right)$, where O is the number of ONUs and Γ_{iq}^{PON} is the traffic coming from ONU$_i$ to ONU$_q$, and Λ denotes the number of wavelengths in the wavelength division multiplexing (WDM) PON.

After calculating the computation and communication delay components, we proceed to compute the total average response time Υ_i of MU i as follows[6]:

$$\Upsilon_i = (1 - \beta_i) \cdot \underbrace{\Delta_{\text{MU}_i}}_{\text{local response time } D_{L,i}^r} + \beta_i \cdot \underbrace{\left(\Theta_i^{\text{UL}} + \left(1 - \alpha_j\right) \Delta_{\text{MEC}_j} + \alpha_j \left(\Theta^{\text{PON}} + \Delta_c\right) \right)}_{\text{nonlocal response time } D_{\text{NL},i}^r} \tag{5.20}$$

where the terms denoted by Δ and Θ represent the latency components of computation and communication, respectively. Note that the communication-induced latency terms Θ_i^{UL} and Θ^{PON} depend on the offloading probabilities β_i and α_j, respectively. More specifically, if MUs decide to offload a large portion of their incoming tasks to the edge servers, the average task transmission delay in the uplink as well as the waiting times in the edge server may increase significantly. On the other hand, if the edge servers also decide to further offload a large portion of their tasks arriving from MUs and fixed subscribers to the remote cloud, the backhaul upstream delay as well as waiting delay at the cloud servers may increase as a result. Therefore, in order for the MUs to benefit from the powerful computational capabilities of the edge/remote servers and experience a low response time, it is important for both device and edge-server schedulers to optimally adjust their offloading probabilities.

5.3.2 Average Energy Consumption per Task

In the following, assuming that the underlying networking and computing components of our considered architecture shown in Figure 5.1 (e.g. BSs, APs, ONUs, and edge servers) are consistently powered with sufficiently available, low-cost energy

---

6 Similar to Chen et al. (2016), Guo and Liu (2018), Liu et al. (2018), and Guo et al. (2016), we neglect the time overhead for sending the computation result back to the MUs due to the fact that for many applications (e.g. face/object recognition) the size of the computation result is generally smaller than that of the computation input data.

resources, we restrict our attention to the energy consumption of MUs only. Similar to Xiao and Krunz (2018), we model the power consumption of MU i's CPU as κf_i^3, where κ is the effective switched capacitance related to the chip architecture (Wang et al., 2016). The energy consumption per CPU cycle is thus equal to κf_i^2, as f_i represents the number of CPU cycles per second. The average energy consumption E_i^l for local execution of a task at MU i is then given by Xiao and Krunz (2018)

$$E_i^l = \kappa \cdot f_i^2 \cdot D_i^l. \tag{5.21}$$

Recall from above that an incoming task at MU i is either executed locally with probability $(1 - \beta_i)$ or it is offloaded for nonlocal execution with probability β_i. The energy consumption, E_i^o, of MU i for offloading an incoming task is given by

$$E_i^o = E_i^{\text{UL}} + E_i^{\text{DL}} \tag{5.22}$$

where E_i^{UL} and E_i^{DL} are the average energy consumptions of MU i to offload an incoming task in the uplink direction and receive its output in the downlink direction, respectively. In the uplink, E_i^{UL} is calculated as follows:

$$E_i^{\text{UL}} = \left(k_1^{\text{tx}} + k_2^{\text{tx}} \cdot p_i^{\text{tx}}\right) \cdot \Theta_i^{\text{UL}} \tag{5.23}$$

whereby k_1^{tx} represents the static power consumption for having the radio frequency (RF) transmission circuitries switched on and k_2^{tx} measures the linear increase of the transmitter power consumption with radiated power p_i^{tx}. In the downlink, E_i^{DL} of MU i is estimated by

$$E_i^{\text{DL}} = \left(k_1^{\text{rx}} + k_2^{\text{rx}} \cdot r_i^{\text{DL}}\right) \cdot \Theta_i^{\text{DL}} \tag{5.24}$$

where k_1^{rx} represents the extra power consumption for having the receiver circuit switched on, k_2^{rx} (measured in W/Mbps) is the power consumption per Mbps in the downlink direction, and r_i^{DL} is the downlink rate, which is given by

$$r_i^{\text{DL}} = P_{\text{temp}}^{\text{MU}} \cdot r^{\text{WiFi}} + (1 - P_{\text{temp}}^{\text{MU}}) \cdot r^{\text{LTE}} \tag{5.25}$$

where r^{WiFi} and r^{LTE} are the average transmission rates of the WiFi access point and LTE BS, respectively. Further, we note that the transmission time, Θ_i^{DL}, of the task output in the downlink direction is estimated by B_i^o/r_i^{DL}, where B_i^o is the task output size. Note that unlike Θ_i^{UL}, Θ_i^{DL} does not depend on the offloading probability β_i. The average energy consumption E_i (for either executing a task locally or transmitting its input data to an edge server) of MU i is then estimated as

$$E_i = \left(1 - \beta_i\right) E_i^l + \beta_i E_i^o. \tag{5.26}$$

By substituting Eqs. (5.21) and (5.22) into Eq. (5.26), we have

$$E_i = \left(1 - \beta_i\right) \left(\kappa \cdot f_i^2 \cdot D_i^l\right) + \beta_i [\left(k_1^{\text{tx}} + k_2^{\text{tx}} \cdot p_i^{\text{tx}}\right) \cdot \Theta_i^{\text{UL}} + \left(k_1^{\text{rx}} + k_2^{\text{rx}} \cdot r_i^{\text{DL}}\right) \cdot \Theta_i^{\text{DL}}]. \tag{5.27}$$

5.4 Energy-Delay Trade-off via Self-Organization

According to our analysis above, an improved QoE is only achieved when an optimal setting of the offloading probabilities is done at both device and edge server sides. Any deviation from this optimal setting may result in performance deterioration. Due to the inherent time-varying nature of the network state, which is a direct consequence of user mobility and traffic fluctuation, such an optimal setting may not be obtained and maintained easily. To cope with this issue, we enable the MUs with self-awareness such that after local measurements, they achieve a reduced average response time and energy consumption by tuning their local parameters only.

In the following, we develop a bicriteria optimization framework to enable MUs to use their local information and minimize the response time as well as their energy consumption by dynamically adjusting their offloading probability as well as CPU clock frequency using the aforementioned DVS technique. For notational simplicity, we consider a tagged user and drop the subscript i hereafter. Similar to Wang et al. (2016) and Guo et al. (2016), we assume that the CPU clock frequency f of the tagged MU is restricted to a continuous interval of $[f_{\min}, f_{\max}]$. We formulate the bicriteria energy-delay self-organization problem as follows:

$$(\mathcal{P}_1): \quad \min_{f,\beta} \quad \Upsilon(f,\beta), E(f,\beta) \tag{5.28a}$$

$$\text{s.t.} \quad f_{\min} \leq f \leq f_{\max} \tag{5.28b}$$

$$0 \leq \beta \leq 1 \tag{5.28c}$$

where Υ and E are given in Eqs. (5.20) and (5.27), respectively. To assess the developed model and characterize the trade-off between the two objectives of the formulation above, we apply the Pareto front analysis. To obtain the Pareto front solutions, a common approach is to transform the original problem into an optimization problem by transferring one of the objectives into the constraints and solving it iteratively. In doing so, the problem $(\mathcal{P}_1)$ is transformed into a single-objective nonlinear optimization problem as follows:

$$(\mathcal{P}_2): \quad \min_{f,\beta} \quad \Upsilon(f,\beta) \tag{5.29a}$$

$$\text{s.t.} \quad f_{\min} \leq f \leq f_{\max} \tag{5.29b}$$

$$0 \leq \beta \leq 1 \tag{5.29c}$$

$$E(\beta,f) \leq \mathcal{E}_{\text{thr}} \tag{5.29d}$$

where $\mathcal{E}_{\text{thr}}$ denotes the given energy budget not to be exceeded (see constraint (5.29d)).

Lemma 5.2 *Problem (5.29) is a convex optimization problem.*

Proof: $\Upsilon(\beta, f)$ is a continuous twice-differentiable convex function of f and β, which can be verified by the fact that its Hessian matrix is positive definite. Besides, constraints 5.29b and 5.29d are affine functions of f and β, respectively; and constraint 5.29d is a convex function of f and β. Therefore, the feasible set of the problem is a convex set. $\qquad\square$

Lemma 5.3 *Necessary condition for optimality: The optimal solution (f^*, β^*) of problem $(\mathcal{P}_2)$ satisfies the following equation*:

$$f^* = \mathcal{L}(\beta^*) = \min\{\mathcal{G}(\beta), f_{\max}\}, \qquad \forall \beta \in [0, \beta_{\max}], \tag{5.30}$$

where

$$\mathcal{G}(\beta) = \sqrt{\frac{\mathcal{E}_{\text{thr}} - \beta[(k_1^{\text{tx}} + k_2^{\text{tx}} p_t)\Theta^{\text{UL}} + (k_1^{\text{rx}} + k_2^{\text{rx}} r^{\text{DL}})\Theta^{\text{DL}}]}{(1 - \beta)\kappa D^l}}, \tag{5.31}$$

and $\beta_{\max}$ is obtained by solving the following equation:

$$\mathcal{G}(\beta) - f_{\min} = 0. \tag{5.32}$$

Proof: First, we show that $f = \mathcal{L}(\beta)$ (given by Eq. (5.30)) determines the upper limit of CPU clock frequency f for a given β. In doing so, we take the energy constraint given by constraint (5.29d) and calculate f as a function of β for a fixed $\mathcal{E}_{\text{thr}}$ as follows:

$$f \leq \mathcal{G}(\beta). \tag{5.33}$$

Taking into account constraint (5.29d), inequality (5.33) becomes:

$$f \leq \overbrace{\min\{\mathcal{G}(\beta), f_{\max}\}}^{\mathcal{L}(\beta)}. \tag{5.34}$$

Clearly, $f = \mathcal{L}(\beta)$ gives the upper limit of f for a given β (see Figure 5.3).

Next, we prove Lemma 5.3 by contradiction. Let (β_1, f_1) satisfy $f_1 = \mathcal{L}(\beta_1)$. Assume $(\beta_1, f_2), \forall f_2 < f_1$, achieves a smaller response time, thus:

$$\Upsilon(\beta_1, f_2) < \Upsilon(\beta_1, f_1). \tag{5.35}$$

Obviously, since $f = \mathcal{L}(\beta)$ gives the upper limit of the CPU clock frequency f for a given (fixed) offloading probability β, an MU can experience a smaller response time by increasing its CPU clock frequency from f_2 to f_1. This happens in light

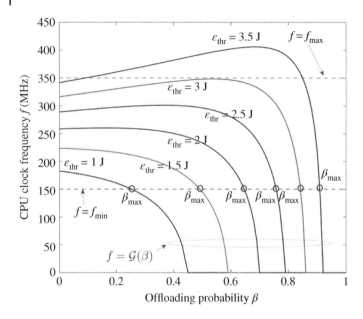

Figure 5.3 Illustration of the search space for problem (P_2) for different values of $\mathcal{E}_{thr}$. Source: Ebrahimzadeh and Maier (2020). © 2020 IEEE.

of the fact that for a fixed offloading probability β_1, as CPU clock frequency f increases, the first term on the right-hand side of Eq. (5.20) decreases, whereas the second term remains unchanged. Hence, $\Upsilon(\beta_1, f_1) < \Upsilon(\beta_1, f_2)$, which is in contradiction with (5.35). We also note that for a given energy budget $\mathcal{E}_{thr}$, the maximum offloading probability β_{max} is obtained by solving $f_{min} = \mathcal{G}(\beta)$, as illustrated in Figure 5.3. Therefore, the optimal solution (f^*, β^*) of problem (P_2) satisfies Eq. (5.30). This completes the proof. □

Since Lemmas 5.2 and 5.3 hold, we use a standard constrained convex optimization approach and obtain the optimal solution (f^*, β^*) of problem (P_2) for a given MU using only its local information/parameters, as described in Algorithm 6, where ϵ_2 is the minimum step size for searching the optimal solution (β^*, f^*), M is a big number, and ϵ_1 is a small number. In Algorithm 6, the optimal offloading probability β^* is obtained via a one-dimensional search method, whereas the optimal CPU clock frequency f^* is calculated accordingly using the closed-form relation given by Eq. (5.30). It is worthwhile to mention that the nonlocal processing latency term, D_{NL}^r, in Eq. (5.20) can be calculated using only local information as follows:

$$D_{NL}^r = \frac{\hat{\Upsilon} - (1 - \beta) \cdot D_L^r - \beta \cdot \hat{\Theta}^{UL}}{\beta} \tag{5.36}$$

Algorithm 6 Joint Offloading and DVS Procedure

Input: $\mathcal{E}_{thr}, f_{min}, f_{max}$, energy and task parameters
Output: β^* and f^*
1: **Initialize:** $\beta_0 \leftarrow 0$ and $f_0 \leftarrow \mathcal{L}(0)$
2: Solve Eq. (5.32) and obtain β_{max}
3: $\Delta \leftarrow M$
4: **while** $(\Delta > \epsilon_1)$ & $(\beta_0 \leq \beta_{max})$ **do**
5: $\quad \beta_1 \leftarrow \beta_0 + \epsilon_2$
6: $\quad f_1 \leftarrow \mathcal{L}(\beta_1)$ using Eq. (5.30)
7: $\quad \Delta \leftarrow \Upsilon(f_0, \beta_0) - \Upsilon(f_1, \beta_1)$ using Eq. (5.20)
8: $\quad$ **if** $\Delta > 0$ **then**
9: $\quad\quad \beta^* \leftarrow \beta_1$ and $f^* \leftarrow f_1$
10: $\quad$ **else**
11: $\quad\quad \beta^* \leftarrow \beta_0$ and $f^* \leftarrow f_0$
12: $\quad$ **end if**
13: $\quad \beta_0 \leftarrow \beta_1$
14: **end while**
15: **return** β^*, f^*

Source: Ebrahimzadeh and Maier (2020). © 2020 IEEE.

where $\hat{\Upsilon}$ and $\hat{\Theta}^{UL}$ can be obtained via measurements. Given that the number of arithmetic operations within each iteration is upper-bounded, the complexity of Algorithm 6 is $\mathcal{O}(K)$, where $K = \frac{\beta_{max}}{\epsilon_2}$ is the number of iterations for searching the optimal solution.

We note that once the optimal setting of the offloading probability and CPU clock frequency is achieved for a given MU, any change in the network state may result in a degraded energy-delay performance. The optimal setting at the servers side is obtained and maintained by solving a centralized optimization problem in the backhaul. In contrast, at the MU side, the optimal setting is achieved and then maintained by periodically solving the problem ($\mathcal{P}_2$) using the local parameters and latency measurements, as explained above. This, as a result, mandates the need for designing an efficient mechanism that periodically updates the optimal setting to ensure that the desired energy-delay performance is maintained.[7]

5.5 Results

The following numerical results were obtained by using the LTE-A and FiWi network and traffic parameter settings listed in Table 5.1, which are consistent

---

[7] Investigating efficient methods of maintaining optimal setting at the MU side while taking into account operational expenditure (OPEX)/capital expenditure (CAPEX) considerations is an interesting topic to pursue as future work.

Table 5.1 MEC-enabled FiWi enhanced HetNet parameters and default values.

Parameter	Value	Parameter	Value
Traffic model parameters			
L_m	1500 Bytes	λ_B	30 packet/s
α_{PON}	100	$\overline{L}, \varsigma_L^2$	1500 Bytes, 0
Backhaul EPON			
l_{PON}	20 km	c_{PON}	10 Gbps
N_{ONU}	$\{12, 16, 20, 24\}$	Λ	1
WiFi parameters			
DIFS	34 μs	SIFS	16 μs
PHY Header	20 μs	W_0, H	16 slots, 6
ϵ	9 μs	RTS	20 bytes
CTS	14 bytes	ACK	14 bytes
r in WMN	300 Mbps	ONU-AP radius	15 m
LTE-A parameters			
p^{tx}	100 mW	ω	5 MHz
$\overline{\omega}_0^2$	−100 dBm	k_1^{tx}	0.4 W
k_2^{tx}	18	ONU-BS radius	50 m
p^{rx}	200 mW	k_1^{rx}	0.4 W
k_2^{rx}	2.86 W/Mbps		
Task and edge/cloud server parameters			
λ_{MU}	25 task/min	f_i	$[150, 450]$ MHz
λ_0^e	30 task/min	f_j^e	1.44 GHz
λ_0^c	240 task/min	s	6
f^c	1.44 GHz	B^l	66 KB
D^l, D_0^e, D_0^c	300 Mcycles	κ	10^{-26}
ϵ_1, ϵ_2	10^{-3}	M	10^2
B^o	1 KB		

with those in Beyranvand et al. (2017), Aurzada et al. (2014), Wang et al. (2016), Guo et al. (2016), and Miettinen and Nurminen (2010). In our considered scenario, 50 MUs are scattered randomly within the range of 50 m from each ONU-BS. Besides, we consider four MUs within the coverage area of each ONU-AP. In the cellular access mode, we set the channel gain to $G_{i,k} = d_{i,k}^{-\xi}$ between MU i and BS k, where $d_{i,k}$ is the distance between MU i and BS k, and $\xi = 4$ is the path loss factor. Further, we set $\beta_i = \beta (\forall i = 1, 2, \dots)$ and $\alpha_j = \alpha (\forall j = 1, 2, \dots)$. A portion ζ of the number of MUs that reside within the coverage area of an ONU-AP or cellular BS is involved in task offloading, while the remaining portion $(1 - \zeta)$ generates conventional Poisson H2H traffic at mean packet rate λ_B. Background traffic coming from ONUs with attached fixed (wired) subscribers is set to $\alpha_{PON} \cdot \lambda_B$, where $\alpha_{PON} \geq 1$ is a traffic scaling factor for fixed subscribers that are directly connected to the backhaul EPON. Also, the user mobility parameters in our simulations are tuned such that the WiFi connection and interconnection times fit a truncated Pareto distribution with $P_{temp}^{MU} = 73.18\%$, which is compliant with the measurements in Beyranvand et al. (2017).

First, we consider the edge computing-only scenario with α being set to zero. Figure 5.4a,b depict the energy-delay performance of MEC-assisted partial offloading. The average response time vs. offloading probability β for different values of ζ is shown in Figure 5.4a. The results indicate that the average response time is a convex function of β. For $\zeta = 20\%$, setting $\beta = 0.66$ leads to an 82% reduction of the average response time compared to the fully local computing scheme (i.e. $\beta = 0$). We note that the optimal value of β largely depends on ζ. More specifically, as ζ

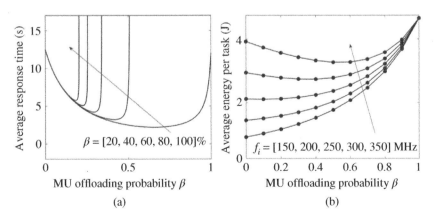

Figure 5.4 (a) Average response time vs. mobile user (MU) offloading probability β for different values of ζ ($\alpha = 0$ and f_i=150 MHz); (b) average energy per task vs. MU offloading probability β for different values of local clock frequency f_i ($\zeta = 20\%$). Source: Ebrahimzadeh and Maier (2020). © 2020 IEEE.

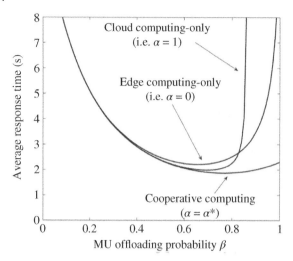

Figure 5.5 Comparison of average response time performance of edge-only, cloud-only, and cooperative computing ($\zeta = 20\%$). Source: Ebrahimzadeh and Maier (2020). © 2020 IEEE.

increases, the optimal value of β decreases, as shown in Figure 5.4a. Figure 5.4b depicts the average energy consumption per task vs. β for different values of f_i. The bottom curves in Figures 5.4a,b highlight the trade-off an MU can make between the average response time and energy per task for $\zeta = 20\%$. We also observe from Figure 5.4b that for larger values of f_i, partial offloading not only reduces the average response time but it also helps MUs reduce their energy consumption.

Next, we examine the performance gains obtained from our proposed trilateral device-edge-cloud cooperative computing. Figure 5.5 depicts the average response time vs. β for the following three different scenarios: (i) edge-only ($\alpha = 0$), (ii) cloud-only ($\alpha = 1$), and (iii) cooperative computing ($\alpha = \alpha^*$), where α^* denotes the optimal value of α set by the network operator to minimize the average task execution time experienced by the edge servers. Figure 5.5 shows that the proposed cooperative computing scheme yields a better delay performance compared to either the edge-only or cloud-only scheme, especially for $\beta > 0.64$. While the edge- and cloud-only schemes may both pose a longer response time due to an excessive queueing delay for large values of β, the trilateral cooperation between the CPU, edge server, and remote cloud yields a reduced response time by setting β and α to their optimal values (see bottom curve in Figure 5.5).

Figure 5.6 shows the impact of edge server offloading probability α on the delay performance of our cooperative computing scheme. For $f_i = 250$ MHz, when MUs operate in the full offloading mode (i.e. $\beta = 1$), the delay of the cooperative computing scheme equals 2.28 seconds by setting $\alpha = 0.68$, compared to ~ 12 seconds of the edge- or cloud-only schemes, which translates into an 81% reduction of the average response time. More interestingly, the reduction is achieved by appropriately setting the edge server offloading probability α and without incurring any

Figure 5.6 Average response time vs. edge-server offloading probability α for different values of β and f_i ($\zeta = 20\%$). Source: Ebrahimzadeh and Maier (2020). © 2020 IEEE.

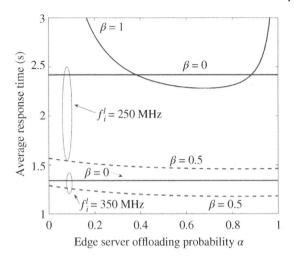

Figure 5.7 Average response time vs. edge-server offloading probability α for different values of ζ ($N_{ONU} = 12$ and $\beta = \beta^*$). Source: Ebrahimzadeh and Maier (2020). © 2020 IEEE.

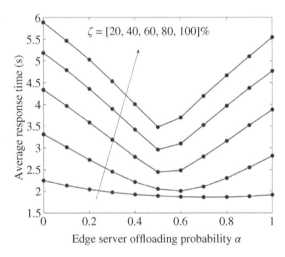

additional energy per task, which remains unchanged for a given β. We note that the average response time as well as the energy per task can be further reduced by setting $\beta = 0.5$. Moreover, for $f_i = 350$ MHz, the minimum response time is achieved by setting $\beta = 0.5$ and $\alpha = 0.86$. In doing so, setting $\beta = 0.5$ also yields a near-optimal energy performance according to Figure 5.4b.

We proceed by discussing the results of the average response time vs. α for different values of ζ in Figure 5.7. We observe that the beneficial impact of edge-cloud cooperation through the backhaul on the average response time becomes even more pronounced for larger values of ζ. Specifically, we observe that the average response time decreases from 5.89 seconds for $\alpha = 0$ to 3.48 seconds for

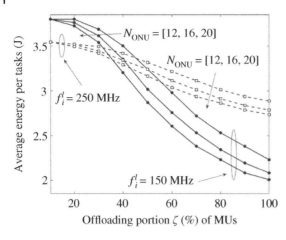

Figure 5.8 Average energy per task vs. ζ for different values of local clock frequency f_i and number N_{ONU} of optical network units (ONUs) ($\alpha = \alpha^*$ and $\beta = \beta^*$). Source: Ebrahimzadeh and Maier (2020). © 2020 IEEE.

$\alpha = \alpha^* = 0.5$ and $\zeta = 100\%$, as opposed to only a slight decrease from 2.25 seconds for $\alpha = 0$ to 1.87 seconds for $\alpha = \alpha^* = 0.7$ and $\zeta = 20\%$.

Next, we examine the energy performance of our proposed cooperative offloading scheme in Figure 5.8, where α and β are set to their optimal values of α^* and β^*. For a given f_i and an increasing ζ or N_{ONU}, we observe a generally decreasing trend in the energy consumption. This occurs because for a larger ζ or N_{ONU}, the optimal delay performance is achieved by relying more on local rather than nonlocal computing resources by setting β to smaller values, which in turn results in a reduced energy consumption (see Figure 5.4b). Importantly, we also observe that while increasing the local clock frequency f_i always leads to a decreased average response time (see Figure 5.6), it may not necessarily increase the energy consumption. For instance, we observe that the energy consumption for $f_i = 250$ MHz, unexpectedly, is lower than that of $f_i = 150$ MHz, provided that $\zeta < 45\%$. This is because the delay-optimal setting β^* in the former is smaller than that of the latter one, thus revealing a better energy performance (see also Figure 5.4b). Note, however, that for $\zeta > 45\%$, setting $f_i = 150$ MHz can achieve an energy saving of up to 38% compared to $f_i = 250$ MHz.

Next, we present the cumulative distribution function (CDF), $F_{\Upsilon_i}(t)$, of the response time Υ_i (i.e. $\Pr\left(\Upsilon_i \leq t\right)$) in Figure 5.9 for different values of ζ. We find that our proposed cooperative computing scheme ensures a lower bound probability of 80% that an incoming task is executed (either locally or nonlocally) and returned to the MU within 5.5 seconds. The average response time vs. α for different values of N_{ONU} is depicted in Figure 5.10, where we evaluate the delay performance of our proposed edge-cloud cooperation through the backhaul. Interestingly, we find that by doubling the number N_{ONU} of ONUs from 12 to 24, the average response time of MUs only increases from 1.88 to 1.95 seconds, provided that an optimal setting of both α and β is carried out.

Figure 5.9 Cumulative distribution function (CDF) of response time for different values of ζ ($\alpha = \alpha^*$ and $\beta = \beta^*$). Source: Ebrahimzadeh and Maier (2020). © 2020 IEEE.

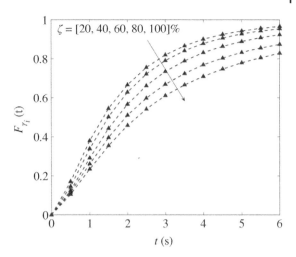

Figure 5.10 Average response time vs. edge server offloading probability α for different values of N_{ONU} ($\beta = \beta^*$, $\zeta = 20\%$). Source: Ebrahimzadeh and Maier (2020). © 2020 IEEE.

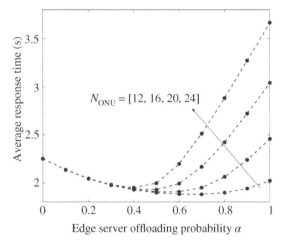

Figure 5.11 illustrates the Pareto frontier solutions of the energy-delay trade-off an MU can choose from via dynamic reconfiguration using our proposed self-organization scheme for different values of D^l. For a given energy constraint $\mathcal{E}_{\text{thr}}$, which is determined by the decision maker according to a given energy budget of the battery as well as delay requirement of incoming tasks, an MU can optimally adjust its offloading probability and CPU clock frequency using only its local information to achieve the desired energy-delay performance. For instance, Figure 5.11 shows that for $D^l = 300$ Mcycles, by increasing $\mathcal{E}_{\text{thr}}$ from 0.7 to 1.8 J an 84% reduction of the average response time is achieved. We also observe from Figure 5.11 that any further increase of the energy budget may not lead to a significant reduction of the average response time, especially for $D^l = 100$

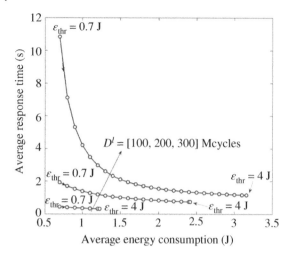

Figure 5.11 Pareto front solutions of self-organization problem (P_2) for $D^l \in [100, 200, 300]$ Mcycles (the value of energy constraint $\mathcal{E}_{\text{thr}}$ increases along the arrow shown on each curve). Source: Ebrahimzadeh and Maier (2020). © 2020 IEEE.

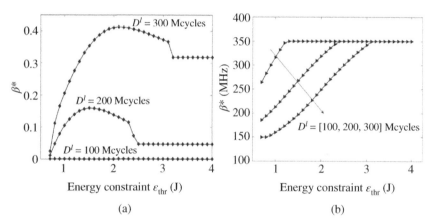

Figure 5.12 (a) Optimal offloading probability β^* vs. energy constraint $\mathcal{E}_{\text{thr}}$ and (b) optimal central processing unit (CPU) clock frequency f^* vs. energy constraint $\mathcal{E}_{\text{thr}}$ for different values of $D^l = [100, 200, 300]$ Mcycles. Source: Ebrahimzadeh and Maier (2020). © 2020 IEEE.

and 200 Mcycles. Moreover, the results of the optimal offloading probability β^* and CPU clock frequency f^* vs. the energy constraint $\mathcal{E}_{\text{thr}}$ for different values of D^l are shown in Figure 5.12. Interestingly, Figure 5.12a,b, along with Figure 5.11, illustrate the impact of increasing D^l and $\mathcal{E}_{\text{thr}}$ on the optimal decision made by the MU. For instance, we observe that for $D^l = 100$ Mcycles, offloading does not have any benefit in terms of average response time, thus $\beta^* = 0, \forall \mathcal{E}_{\text{thr}} \in [0.7, 3]$ J

Figure 5.13 Average uplink delay vs. human-to-human (H2H) background traffic for different values of energy budget $\mathcal{E}_{\text{thr}} = [1, 2, 3, 4]$ J ($\beta = \beta^*$ and $D^l = 300$ Mcycle). Source: Ebrahimzadeh and Maier (2020). © 2020 IEEE.

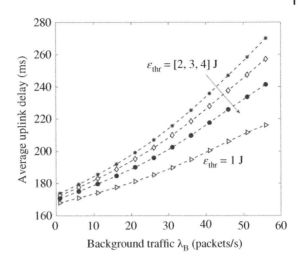

(see Figure 5.12a). Instead, MUs can reduce their response time by increasing f (see Figure 5.12b).

Finally, the average uplink delay vs. H2H background traffic rate λ_B is illustrated in Figure 5.13, which gives insights into how an increasing λ_B contributes to an increased uplink delay. More importantly, according to Figure 5.13, increasing the energy budget $\mathcal{E}_{\text{thr}}$ from 1 to 2 J results in an increased average uplink delay, whereas a further increase of $\mathcal{E}_{\text{thr}}$ from 2 to 4 J leads to a decreased average uplink delay. The reason for this can be inferred from Figure 5.12a, which gives us insights into how the optimal offloading probability β^* is obtained for different values of $\mathcal{E}_{\text{thr}} = [1, 2, 3, 4]$ J.

5.6 Conclusions

This chapter studied the cooperative computation offloading in MEC-enabled FiWi enhanced HetNets from both network architecture and offlading mechanism design perspectives. Beside the design of reliable low-latency MEC-enabled FiWi enhanced LTE-A HetNets, we presented a simple but efficient offloading strategy that leverages trilateral cooperation among device, edge server, and remote cloud. We developed an analytical framework to estimate the average response time and energy consumption of MUs in a FiWi-based MEC-enabled network infrastructure. Our results demonstrate the superior performance of the proposed cooperative computing scheme compared to edge- or cloud-only schemes. Further, we showed that by optimally setting the offloading probabilities, MUs can achieve a reduction of the average response time of up to 81%. In order to cope with the incurred complexity, we also designed a self-organization based mechanism, which enables an MU, using local information, to make suitable energy-delay trade-offs and jointly minimize the average execution time and energy consumption by dynamically adjusting the offloading probability and its local CPU clock frequency using the DVS technique.

6

Decentralization via Blockchain

6.1 Introduction

The Internet has been constantly evolving from the mobile Internet to the emerging Internet of Things (IoT) and future Tactile Internet. Similarly, the capabilities of 5G networks have extended far beyond those of previous generations of mobile communication. Beside 1000-fold gains in area capacity, 10 Gb/s peak data rates, and connections for at least 100 billion devices, an important aspect of the 5G and beyond vision is *decentralization*. While 2G–3G–4G cellular networks were built under the design premise of having complete control at the infrastructure side, 5G and beyond systems may drop this design assumption and evolve the cell-centric architecture into a more device-centric one. While there is a significant overlap of design objectives among 5G, IoT, and the Tactile Internet – most notably ultra-reliable and low-latency communications (URLLC) – each one of them exhibits unique characteristics in terms of underlying communications paradigms and enabling end-devices (Maier et al., 2016).

Today's Internet is ushering in a new era. While the first generation of digital revolution gave rise to the Internet of information, the second generation – powered by decentralized blockchain technology – is bringing us the Internet of value, a true peer-to-peer platform that has the potential to go far beyond digital currencies and record virtually everything of value to humankind in a distributed fashion without powerful intermediaries (Tapscott and Tapscott, 2016). Some refer to decentralized blockchain technology as the "alchemy of the twenty-first century" since it may leverage end-user equipment for converting computing into digital gold. Arguably more importantly, though, according to Don and Alex Tapscott, the blockchain technology enables trusted collaboration that can start to change the way wealth is distributed as people can share more fully in the wealth they create. As a result, decentralized blockchain technology helps create platforms

Toward 6G: A New Era of Convergence, First Edition. Amin Ebrahimzadeh and Martin Maier.
© 2021 The Institute of Electrical and Electronics Engineers, Inc.
Published 2021 by John Wiley & Sons, Inc.

for distributed capitalism and a more inclusive economy, which works best when it works for everyone as the foundation for prosperity. Furthermore, the authors of Saad et al. (2020) pointed out the important role of blockchain and distributed ledger technology (DLT) applications as a next-generation of distributed sensing services for 6G driving applications whose need for connectivity will require a synergistic mix of URLLC and massive machine type communications (mMTC) to guarantee low-latency, reliable connectivity, and scalability. Furthermore, blockchains and smart contracts can improve the security of a wide range of businesses by ensuring that data cannot be damaged, stolen, or lost. In Salman et al. (2019), the authors presented a comprehensive survey on the utilization of blockchain technologies to provide distributed security services. These services include entity authentication, confidentiality, privacy, provenance, and integrity assurances.

The fundamental concepts and potential of blockchain technologies for society and industry in general have been described comprehensively in various existent tutorials, e.g. Beck (2018). In particular, there has been a growing interest in adapting blockchain to the specific needs of the IoT in order to develop a variety of blockchain-based Internet of things (BIoT) applications, ranging from smart cities and Industry 4.0 to financial transactions and farming, among others Novo (2018). Toward this end, the authors of Fernández-Caramés and Fraga-Lamas (2018) pointed out the important role of smart contracts, which are defined as pieces of self-sufficient decentralized code that are executed autonomously when certain conditions are met, whereby *Ethereum* was argued to be the most popular blockchain-based platform for running smart contracts. The use of Ethereum allows users to write and run their own code on top of the network. By updating the code, users are able to modify the behavior of IoT devices for simplified maintenance and error correction. Beside well-known BIoT problems such as hosting a blockchain on resource-constrained IoT devices, low transaction rates, and long block creation times, the authors of Fernández-Caramés and Fraga-Lamas (2018) identified several significant challenges beyond early BIoT developments and deployments that will need further investigation. Apart from technological challenges, e.g. access control and security, the authors concluded that shaping the regulatory environment, e.g. *decentralized ownership*, is one of the biggest issues to unleash the potential of BIoT for its broader use.

Recently, initial studies have begun to address some of the aforementioned shortcomings of BIoT. In Pan et al. (2019), resource-constrained IoT devices were released from computationally intensive tasks by offloading the mining work (i.e. creating/appending/monitoring blockchain transactions) onto more powerful edge computing resources such as cloudlets. The proposed *EdgeChain* was built on the Ethereum platform and uses smart contracts to monitor and regulate the behavior of IoT devices based on how they act and use resources.

Since all activities are stored in the blockchain, it is inconvenient for malicious nodes to cause serious damage or take off without any evidences. Furthermore, to tackle the critical access control issue of preventing BIoT resource access from unauthorized entities, the authors of Zhang et al. (2019a) exploited the Ethereum smart contract platform to achieve various access control methods. Specifically, gateways were used to act as BIoT service agents for their respective cluster of local resource-constrained IoT devices by storing their blockchain accounts and using them to execute smart contracts on their behalf. The proposed smart contract based framework consists of multiple access control contracts (ACCs). Each ACC maintains a misbehavior list for each BIoT resource, including details and time of the misbehavior as well as the penalty on its subject, e.g. blocking access requests for a certain period of time. Further, in addition to a register contract, the framework involves the so-called *judge contract (JC)*, which implements a certain misbehavior judging method. After receiving the misbehavior reports from the ACCs, the JC determines the penalty on the corresponding subjects and returns the decisions to the ACCs for execution.

Many additional BIoT studies considered Ethereum as the blockchain of choice. For instance, the architectural issues for realizing BIoT services were investigated in greater detail in Liao et al. (2017). In a preliminary study using a smart thing renting service as an example BIoT service, the authors compared the following four different architectural styles based on Ethereum: (i) fully centralized (cloud without blockchain), (ii) pseudo distributed things (physically located in central cloud), (iii) distributed things (directly controlled by smart contract), and (iv) fully distributed. The preliminary results indicate that a fully distributed architecture, where a blockchain endpoint is deployed on the end-user device, is superior in terms of robustness and security. Further, the various perspectives for integrating secure elements in Ethereum transactions were discussed in Urien (2018). To prevent the risks that secret keys for signature are stolen or hacked, the author proposed to use java card secure elements and a so-called *crypto currency smart card (CCSC)*. Two CCSC use cases were discussed. In the first one, the CCSC was integrated in a low-cost BIoT device powered by an Arduino processor, in which sensor data are integrated in Ethereum transactions. The second use case involved the deployment of CCSC in remote APDU call secure (RACS) servers to enable remote and safe digital signatures by using the well-known elliptic curve digital signature algorithm (ECDSA).

Despite the recent progress, the salient features that set Ethereum aside from other blockchains remain to be explored in more depth, including their symbiosis with other emerging key technologies such as artificial intelligence (AI) and robots apart from decentralized edge computing. A question of particular interest hereby is how decentralized blockchain mechanisms may be leveraged to let emerge new hybrid forms of collaboration among individuals, which have not been entertained in the traditional market-oriented economy dominated by firms rather than individuals (Beck, 2018). Of particular interest will be Ethereum's concept of the so-called *decentralized autonomous organization (DAO)*. In fact, in their latest book on how to harness our digital future (McAfee and Brynjolfsson, 2017), Andrew McAfee and Erik Brynjolfsson speak of "The Way of The DAO" that may substitute a technology-enabled crowd for traditional organizations such as companies. Toward this end, we focus on the *Tactile Internet*, which is considered the next leap in the evolution of today's IoT. Recall from Chapter 2 that the IoT with its underlying machine-to-machine (M2M) communications is designed to enable communications among machines without relying on any human involvement. Conversely, the Tactile Internet will bring a new dimension to human-to-machine interaction involving its intrinsic human-in-the-loop (HITL) nature. As a consequence, this enables to create new immersive applications and extend the capabilities of the human through the Internet via a human-centric design approach, i.e. augmenting instead of automating away the human (Maier et al., 2018). This chapter aims at addressing the open research challenges outlined above.

The remainder of the chapter is organized as follows. In Section 6.2, after elaborating on the commonalities of and specific differences between Ethereum and Bitcoin blockchains, we explain DAO in more detail. Section 6.3 reviews recent progress and open challenges of the emerging BIoT and edge computing. Section 6.4 discusses the potential role of Ethereum and in particular the DAO in helping decentralize the Tactile Internet. In Section 6.5, we explore possibilities to extend the BIoT framework of JC to nudge contract for enabling the *nudging* of human users in a broader Tactile Internet context. Finally, Section 6.6 concludes the chapter.

6.2 Blockchain Technologies

In this section, we give a brief overview of the basic concepts of blockchain technologies, paying particular attention to the main commonalities and specific differences between Ethereum and Bitcoin. We then introduce the DAO, which represents a salient feature of Ethereum that cannot be found in Bitcoin.

6.2.1 Ethereum vs. Bitcoin Blockchains

Blockchain technologies have been undergoing several iterations as both public organizations and private corporations sought to take advantage of their potential. A typical blockchain network is essentially a distributed database (also known as ledger) comprising records of all transactions or digital events that have been executed by or shared among participating parties. Blockchains may be categorized

Table 6.1 Public vs. private blockchains.

	Public blockchain	Private blockchain
Network type	Fully decentralized	Partially decentralized
Access	Permissionless read/write	Permissioned read/write
User identity	Pseudo-anonymous	Known participants
Consensus mechanism	Proof-of-work/proof-of-stake	Pre-approved participants
Consensus determination	By all miners	By one organization
Immutability	Nearly impossible to tamper	Could be tampered
Purpose	Any decentralized applications	Business applications

into public (i.e. permissionless) and private (i.e. permissioned) networks. In the former category, anyone may join and participate in the blockchain. Conversely, a private blockchain applies certain access control mechanisms to determine who can join the network. A public blockchain is immutable because none of the transactions can be tampered or changed. Also, it is pseudo-anonymous because the identity of those involved in a transaction is represented by an address key in the form of a random string. Table 6.1 highlights the major differences between public and private blockchains, as discussed in further detail below. Note that both Ethereum and Bitcoin are public blockchains.

Figure 6.1 illustrates the main commonalities of and differences between Bitcoin and Ethereum blockchains. The Bitcoin blockchain is predominantly designed to facilitate Bitcoin transactions. It is the world's first fully functional digital currency that is truly decentralized, open source, and censorship resistant. Bitcoin makes use of a cryptographic *proof-of-work (PoW)* consensus mechanism based on the SHA-256 hash function and digital signatures. Achieving consensus provides extreme levels of fault tolerance, ensures zero downtime, and makes data stored on the blockchain forever unchangeable and censorship-resistant in that everyone can see the blockchain history, including any data or messages. There are two different types of actors, whose roles are defined as follows:

- *Regular nodes*: A regular node is a conventional actor, who just has a copy of the blockchain and uses the blockchain network to send or receive Bitcoins.
- *Miners*: A miner is an actor with a particular role, who builds the blockchain through the validation of transactions by creating blocks and submitting them to the blockchain network to be included as blocks. Miners serve as protectors of the network and can operate from anywhere in the world as long as they have sufficient knowledge about the mining process, the hardware and software required to do so, and an Internet connection.

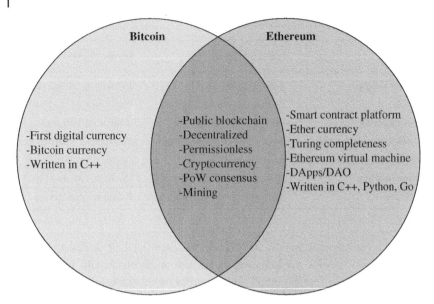

Figure 6.1 Bitcoin and Ethereum blockchains: commonalities and differences.

In the Bitcoin blockchain, a block is mined about every 10 minutes and the block size is limited to 1 MByte. Note that the Bitcoin blockchain is restricted to a rate of seven transactions per second, which renders it unsuitable for high-frequency trading. Other weaknesses of the Bitcoin blockchain include its script language, which offers only a limited number of small instructions and is non-Turing-complete. Furthermore, developing applications using the Bitcoin script language requires advanced skills in programming and cryptography.

Ethereum is currently the second most popular public blockchain after Bitcoin. It has been developed by the Ethereum Foundation, a Swiss nonprofit organization, with contributions from all over the world. Ethereum has its own cryptocurrency called *Ether*, which provides the primary form of liquidity allowing for exchange of value across the network. Ether also provides the mechanism for paying and earning transaction fees that arise from supporting and using the network. Like Bitcoin, Ether has been the subject of speculation witnessing wide fluctuations. Ethereum is well suited for developing *decentralized applications (DApps)* that need to be built quickly and interact efficiently and securely via the blockchain platform. Similar to Bitcoin, Ethereum uses a PoW consensus method for authenticating transactions and proving the achievement of a certain amount of work. The hashing algorithm used by the PoW mechanism is called Ethash. Different from Bitcoin, Ethereum developers expect to replace PoW with a so-called *proof-of-stake (PoS)* consensus. PoS will require Ether miners to hold some amount of Ether, which will be forfeit if the miner attempts to attack the blockchain network. The Ethereum platform is often referred to as a Turing-complete *Ethereum virtual machine (EVM)* built on top of the underlying blockchain. Turing-completeness means that any system or programming language is able to compute anything computable provided it has enough resources. Note that the EVM requires a small amount of fees for executing transactions. These fees are called gas and the required amount of gas depends on the size of a given instruction. The longer the instruction, the more gas is required.

While the Bitcoin blockchain simply contains a list of transactions, Ethereum's basic unit is the *account*. The Ethereum blockchain tracks the state of every account, whereby all state transitions are transfers of value and information between accounts. The account concept is considered an essential component and data model of the Ethereum blockchain since it is vital for a user to interact with the Ethereum network via transactions. Accounts represent the identities of external agents (e.g. human or automated agents, mining nodes). Accounts use public key cryptography to sign each transaction such that the EVM can securely validate the identity of the sender of the transaction.

Beside C++, Ethereum supports several programming languages based on JavaScript and Python, e.g. Solidity, Serpent, Mutan, or Lisp like language (LLL), whereby Solidity is the most popular language for writing smart contracts. A smart contract is an agreement that runs exactly as programmed without any third-party interference. It uses its own arbitrary rules of ownership, transaction formats, and state-transition logic. Each method of a smart contract can be invoked via either a transaction or another method. Smart contracts enable the realization of DApps, which may look exactly the same as conventional applications with regard to application programming interface (API), though the centralized backend services are replaced with smart contracts running on the decentralized Ethereum network without relying on any central servers. Interesting examples of existent DApps include Augur (a decentralized prediction market), Weifund (an open platform for crowdfunding), Golem (supercomputing), and Ethlance (decentralized job market platform), among others. To provide an effective means of communications between DApps, Ethereum uses the *Whisper* peer-to-peer protocol, a fully decentralized middleware for secret messaging and digital cryptography. Whisper supports the creation of confidential communication routes without the need for a trusted third party. It builds on a peer sampling service that takes into account network limitations such as network address translation (NAT) and firewalls. In general, any centralized service may be converted into a DApp by using the Ethereum blockchain.

6.2.2 Ethereum: The DAO

Ethereum made great strides in having its technology accepted as the blockchain standard, when Microsoft Azure started offering it as a service in November 2015. Ethereum was founded by Vitalik Buterin after his request for creating a wider and more general scripting language for the development of DApps that are not limited to cryptocurrencies, a feature that Bitcoin did not have, was rejected by the Bitcoin community (Buterin, 2013). Ethereum enables new forms of economic organization and distributed models of companies, businesses, and ownership, e.g. self-organized holacracies and member-owned cooperatives. Or as Buterin puts it, while most technologies tend to automate workers on the periphery doing unskilled tasks, Ethereum automates away the center. For instance, instead of putting the cab driver out of a job, Ethereum puts Uber out of a job and lets the cab drivers work with the customer directly (before Uber's self-driving cars will eventually wipe out their jobs). Hence, Ethereum does not aim at eliminating jobs so much as it changes the definition of work. In fact, it gave rise to the first DAO built within the Ethereum project. The DAO is an open-source, distributed software that exists "simultaneously nowhere and everywhere," thereby creating a paradigm shift that offers new opportunities to democratize business and enable entrepreneurs of the future to design their own entirely virtual organizations customized to the optimal needs of their mission, vision, and strategy to change the world (McAfee and Brynjolfsson, 2017).

A successful example of deploying the DAO concept for automated smart contract operation is *Storj*, which is a decentralized, secure, private, and encrypted cloud storage platform that may be used as an alternative to centralized storage providers like Dropbox or Google Drive. A DAO may be funded by a group of individuals who cover its basic costs, giving the funders voting rights rather than any kind of ownership or equity shares. This creates an autonomous and transparent system that will continue on the network for as long as it provides a useful service for its customers. DAOs exist as open-source, distributed software that executes smart contracts and works according to specified governance rules and guidelines. Buterin described on the Ethereum Blog the ideal of a DAO as follows: It is an entity that exists on the Internet in an autonomous manner and heavily relies on hiring individuals to execute specific tasks that the automation is unable to do so. AI based agents are fully autonomous, as opposed to a DAO, which still calls for human involvement specifically interacting following a protocol defined by the DAO in order to operate. For illustrating the distinction between a DAO and AI, Figure 6.2 depicts a quadrant chart that classifies DAOs, AI, traditional organizations as well as robots, which been widely deployed in assembly lines among others, with regard to automation and humans involved at their edges and center. We will elaborate on how this particular feature of DAOs (i.e. automation at the

Figure 6.2 Decentralized autonomous organizations (DAOs) vs. artificial intelligence, traditional organizations, and robots (widely deployed in assembly lines, among others): Automation and humans involved at their edges and center. Source: Maier and Ebrahimzadeh (2019). © 2019 IEEE.

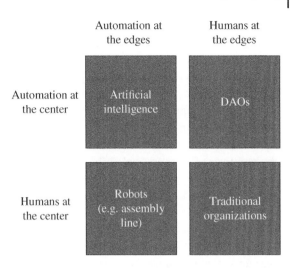

center and humans at the edges) can be exploited for decentralizing the Tactile Internet below in Section 6.4. Toward this end, we also briefly note that according to Buterin even though a DAO is nonprofit, one can make money in a DAO, not by investing into the DAO itself but by participating in its ecosystem, e.g. via *membership* (to be further explored in Section 6.4 in the context of human-agent-robot teamwork [HART] membership).

6.3 Blockchain IoT and Edge Computing

In this section, after defining the integration of BIoT and edge computing, we discuss the motivation of such integration followed by a description of the challenges of integrating blockchain and edge computing.

6.3.1 Blockchain IoT (BIoT): Recent Progress and Related Work

Recall from Section 6.1 that the IoT is designed to enable communications among machines without relying on any human involvement. Thus, its underlying M2M communications is useful for enabling the automation of industrial and other machine-centric processes. The emerging BIoT represents a powerful combination of two massive technologies – blockchain and M2M communications – that allows us to automate complex multi-step IoT processes, e.g. via smart contracts. With the ever-increasing variety of communication protocols between IoT devices, there is a need for transparent yet highly secure and reliable IoT device management systems. This section surveys the state of the art of the emerging BIoT, describing recent progress and open challenges.

The majority of IoT devices are resource constrained, which restricts them to be part of the blockchain network. To cope with these limitations, the author of Novo (2018) proposed a decentralized access management system, where all entities are part of an Ethereum blockchain except for IoT devices as well as so-called *management hub* nodes that request permissions from the blockchain on behalf of the IoT devices belonging to different wireless sensor networks. In addition, entities called *managers* interact with the smart contract hosted at a specific *agent node* in the blockchain in order to define and/or modify the access control policies for the resources of their associated IoT devices. The proof-of-concept implementation evaluated the new system architecture components that are not part of the Ethereum network, i.e. management hub and IoT devices, and demonstrated the feasibility of the proposed access management architecture in terms of latency and scalability. Another interesting Ethereum case study can be found in Aung and Tantidham (2017), which reviews readily available Ethereum blockchain packages for realizing a smart home system according to its smart contract features for handling access control policy, data storage, and data flow management.

Blockchain transactions require public-key encryption operations such as digital signatures. However, not all BIoT devices can support this computationally intensive task. For this reason, in Polyzos and Fotiou (2017), the authors proposed a preliminary design of a *gateway-oriented* approach, where all blockchain related operations are offloaded to a gateway. The authors noted that their approach is compatible with the Ethereum client side architecture.

Due to the massive scale and distributed nature of IoT applications and services, blockchain technology can be exploited to provide a secure, tamper-proof BIoT network. More specifically, the key properties of tamper-resistance and decentralized trust allow us to build a secure authentication and authorization service, which does not have a single point of failure. Toward this end, the authors of Gupta et al. (2018) made a preliminary attempt to develop a security model backed by blockchain that provide confidentiality, integrity, and availability of data transmitted and received by nodes in a BIoT network. The proposed solution encompasses a *blockchain protocol layer* on top of the TCP/IP transport layer and a *blockchain application layer*. The first one comprises a distributed consensus algorithm for BIoT nodes while the latter one defines the IoT security specific transactions and their semantics for the higher protocol layers. To evaluate the feasibility and performance of the proposed layered architecture, BIoT nodes connected in a tree topology were simulated using 1 Gbps Ethernet or 54 Mbps WiFi links. The simulation results showed that the block arrival rate was not affected much by the increased latency and reduced bandwidth, when replacing wired Ethernet with wireless WiFi links, as the block difficulty adjustment adapts dynamically to the network conditions.

Among various low-power wide-area (LPWA) technologies, *long range (LoRa)* wireless radio frequency (RF) is considered one of the most promising enabling technologies for realizing massive IoT deployment. In Özyılmaz and Yurdakul (2017), the authors presented a proof-of-concept demonstrator to enable low-power, resource-constrained LoRa IoT end-devices to access an Ethereum blockchain network via an intermediate gateway, which acts as a full blockchain node. More specifically, a battery-powered IoT end-device sends position data to the LoRa gateway, which in turn forwards it through the standard Go-lang-based Ethereum client *Geth* to the blockchain network using a smart contract. An event-based communication mechanism between the LoRa gateway and a backend application server was implemented as proof-of-concept demonstrator.

One of the fundamental challenges of object identification in IoT stems from the traditional domain name system (DNS). Typically, DNS is managed in centralized modules and thus may cause large-scale failures due to unilateral advanced persistent threat (APT) attacks as well as zone file synchronization delays in larger systems. Clearly, a more robust and distributed name management system is needed that supports the smooth evolution of DNS and renders it more efficient for IoT and the future Internet in general. Toward this end, a decentralized blockchain-based DNS called *DNSLedger* was introduced in Duan et al. (2018). To rebuild the hierarchical structure of DNS, DNSLedger contains two kinds of blockchain: (i) a single *root chain* that stores all the top-level domain (TLD) information and (ii) multiple *TLD chains*, each responsible for the information about its respective domain name. In DNSLedger, servers of domain names act as blockchain nodes, while each TLD chain may select one or more servers to join the root chain. DNSLedger clients may execute common DNS functions such as domain name look-up, application, and modification.

Many of the aforementioned studies considered Ethereum as the blockchain of choice. It was shown that fully distributed Ethereum architectures are able to enhance both robustness and security. Furthermore, a gateway-oriented design approach was often applied to offload computationally intensive tasks from low-power, resource-constrained IoT end-devices onto an intermediate gateway and thus enable them to access the Ethereum blockchain network. Also, it was shown that the block arrival rate does not deteriorate much by the increased latency and reduced bandwidth of WiFi access links.

6.3.2 Blockchain Enabled Edge Computing

One of the critical challenges in cloud computing is the end-to-end responsiveness between the mobile device and an associated cloud. To address this challenge, multi-access edge computing (MEC) is proposed. An MEC entity is a trusted, resource-rich computer or cluster of computers that is well-connected to the

Internet and available for use by nearby mobile devices. According to the European Telecommunications Standards Institute (ETSI), MEC is considered a key emerging technology for next-generation networks. In light of the aforementioned arguments, the integration of blockchain and edge computing into one unified entity becomes a natural trend. On one hand, by incorporating blockchain into the edge computing network, the system can provide reliable access and control of the network, computation, and storage over decentralized nodes. On the other hand, edge computing enables blockchain storage and mining computation from power-limited devices. Furthermore, off-chain storage and off-chain computation at the edges enable scalable storage and computation on the blockchain (Yang et al., 2019). Several recent studies on blockchain and edge computing have been carried out. A blockchain-enabled computation offloading scheme for IoT with edge computing capabilities, called BeCome, was proposed in Xu et al. (2020). The authors of this study aimed at decreasing the task offloading time and energy consumption of edge computing devices, while achieving load balancing and data integrity.

The study in Zhaofeng et al. (2020) proposed a blockchain-based trusted data management scheme called BlockTDM for edge computing to solve the data trust and security problems in an edge computing environment. Specifically, the authors proposed a flexible and configurable blockchain architecture that includes a mutual authentication protocol, flexible consensus, smart contract, block and transaction data management as well as blockchain node management and deployment. The BlockTDM scheme is able to support matrix-based multichannel data segment and isolation for sensitive or privacy data protection. Moreover, the authors designed user-defined sensitive data encryption before the transaction payload is stored in the blockchain system. They implemented a conditional access and decryption query of the protected blockchain data and transactions through an appropriate smart contract. Their analysis and evaluation show that the proposed BlockTDM scheme provides a general, flexible, and configurable blockchain-based paradigm for trusted data management with high credibility. In summary, blockchain-enabled edge computing has become an important concept that leverages decentralized management and distributed services to meet the security, scalability, and performance requirements of next-generation communications networks, as discussed in technically greater detail next.

6.4 Decentralizing the Tactile Internet

In this section, we explore how Ethereum blockchain technologies, in particular the DAO, may be leveraged to decentralize the Tactile Internet as a promising example of future techno-social systems, which at the moment is still debatable in many ways how this would work exactly (Beck, 2018). We search for synergies between the aforementioned HART membership and the complementary strengths of the DAO, AI, and robots (see Figure 6.2) to enable local human-machine coactivity clusters via decentralizing the Tactile Internet. Toward this end, it is important to better understand the merits and limits of AI. Recently, Stanford University launched its *One Hundred Year Study on Artificial Intelligence (AI100)*. In the inaugural report "Artificial Intelligence and Life in 2030," the authors defined AI as a set of computational technologies that are inspired by how people use their brains to sense, learn, reason, and act. They also point out that AI will likely replace tasks rather than jobs in the near term and highlight the importance of *crowdsourcing* of human skills to solve problems that machines alone cannot solve well. As interconnected computing power has spread around the world and useful platforms have been built on top of it, the crowd has become a demonstrably viable and valuable resource. According to McAfee and Brynjolfsson (2017), there are many ways for companies that are squarely at the core of modern capitalism to tap into the expertise of uncredentialed and conventionally inexperienced members of the technology-enabled crowd such as the DAO.

6.4.1 AI-enhanced MEC

First, let us explore the potential of leveraging mobile end-user equipment by partially or fully decentralizing MEC. Recall from above that we introduced the use of AI-enhanced MEC servers at the optical-wireless interface of FiWi-enhanced mobile networks. In a BIoT context, these MEC servers have been used as gateways that are required to act as BIoT service agents to release resource-constrained IoT devices from computation-intensive tasks by offloading blockchain transactions onto more powerful edge computing resources, as discussed in Section 6.1. This design constraint can be relaxed in the Tactile Internet, where user equipment (e.g. state-of-the-art smartphones or the aforementioned user-owned robots) is computationally more resourceful than IoT devices and thus may be exploited for decentralization.

Assuming the network architecture as well as the same default network parameter setting and simulation setup as in Chapter 3, we consider $1 \leq N_{\text{Edge}} \leq 4$ AI-enhanced MEC servers, each associated with eight end-users, whereof $1 \leq N_{\text{PD}} \leq 8$ partially decentralized end users can flexibly control the amount of offloaded tasks by varying their computation offloading probability. The remaining $8 - N_{\text{PD}}$ are fully decentralized end-users that rely on edge computing only (i.e. their computation offloading probability equals 1). Note that for $N_{\text{Edge}} = 4$, all end-users may offload their computation tasks onto an edge node. Conversely,

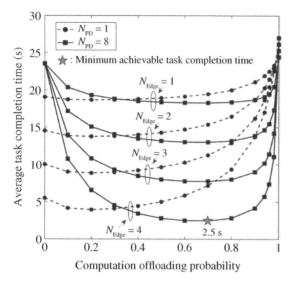

Figure 6.3 Average computational task completion time (in seconds) vs. computation offloading probability for different numbers of partially decentralized endusers (N_{PD}) and artificial intelligence (AI)-enhanced multi-access edge computing (MEC) servers (N_{Edge}).

for $N_{Edge} < 4$, one or more edge nodes are unavailable for computation offloading and their associated endusers fall back on their local computation resources (i.e. fully decentralized). Figure 6.3 shows the average task completion time vs. computation offloading probability of the partially decentralized end users for different N_{Edge} and N_{PD}. We observe from Figure 6.3 that for a given N_{Edge}, increasing N_{PD} (i.e. higher level of decentralization) is effective in reducing the average task completion time. Specifically, for $N_{Edge} = 4$, a high decentralization level ($N_{PD} = 8$) allows end-users to experience a reduction of the average task completion time of up to 89.5% by optimally adjusting their computation offloading probability to 0.7.

Note that in Figure 6.3, the average task completion time is on the order of seconds, ranging from 2.5 to 25 seconds depending on the computation offloading probability. Hence, given Ethereum's transaction limit of 20 transactions/s, the notoriously low transaction rate of blockchain technologies does not pose a significant challenge to the execution of computational tasks and especially physical tasks carried out by robots in the context of the Tactile Internet, as explained in more detail next.

6.4.2 Crowdsourcing

In Chapter 3, we leveraged on self-awareness to introduce the idea of shared use of user- and network-owned robots and developed a self-aware allocation algorithm of physical tasks for HART-centric task coordination. By using our AI-enhanced MEC servers as autonomous agents, we showed in Chapter 2 that delayed force feedback samples coming from teleoperator robots (TORs) may be locally generated and delivered to human operators (HOs) in close proximity. More specifically, we deployed artificial neural network (ANN) to build a forecaster of delayed (or lost) force feedback samples. We showed that by generating the forecast samples at the HO side instead of waiting for the delayed ones, AI-enhanced MEC servers enable HOs to perceive the remote physical task environment in real-time at a 1-ms granularity and thereby experience improved closeness and enhanced safety control therein. Note, however, that the performance of a teleoperation system exploiting sample forecasting largely depends on the accuracy of the forecaster.

In the following, we explore how crowdsourcing helps decrease the completion time of physical tasks in the event of unreliable forecasting of force feedback samples from TORs. Toward realizing DAO in a decentralized Tactile Internet, Ethereum may be used to establish HO–TOR sessions for remote physical task execution, whereby smart contracts help establish/maintain trusted HART membership and allow each HART member to have global knowledge about all participating HOs, TORs, and MEC servers that act as autonomous agents. We assume that an HO remotely executes a given physical task until three out of the recent five haptic feedback samples are misforecast. At this point, the HO immediately stops the teleoperation and informs the agent. The agent assigns the interrupted task to a nearby human (e.g. an available HO) in vicinity of the TOR, who then traverses to the task point and finalizes the physical task. Figure 6.4 depicts the average task completion time vs. probability of sample misforecast for different traverse time T_{traverse} of the nearby human and different ratio of human and robot operational

Figure 6.4 Average physical task completion time (in seconds) vs. probability of sample misforecast for different traverse time $T_{\text{traverse}} \in \{2, 5\}$ seconds of nearby human and different ratio of human and robot operational capabilities $\frac{f_{\text{human}}}{f_{\text{robot}}}$.

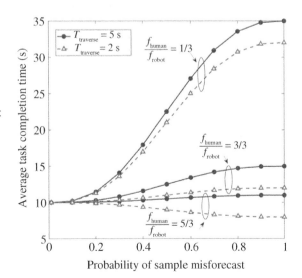

capabilities $\frac{f_{\text{human}}}{f_{\text{robot}}}$, where f_{human} and f_{robot} denote the number of operations per second a human and robot is capable of performing, respectively. We can make several observations from Figure 6.4. Obviously, it is beneficial to select humans with a shorter traverse time, who happen to be closer to the interrupted TOR. We also observe that the ratio $\frac{f_{\text{human}}}{f_{\text{robot}}}$ has a significant impact on the average task completion time. Clearly, for a ratio of smaller than 1 (i.e. 1/3), the human assistance is less useful since it takes him/her more time to complete the physical task. Conversely, for a ratio of equal to 1 (i.e. 3/3) and especially larger than 1 (i.e. 5/3), crowdsourcing pays off by making use of the superior operational capabilities of the human. Whether humans or robots are better suited to perform a physical task certainly depends on its nature. However, for a given physical task, an interesting approach to benefit from the assistance of even uncredentialed and inexperienced crowd members of the DAO may be to enhance the capabilities of humans by means of *nudging*, as explained next.

6.5 Nudging: From Judge Contract to Nudge Contract

6.5.1 Cognitive Assistance: From AI to Intelligence Amplification (IA)

A widely studied approach to increase the usefulness of crowdsourcing has been edge computing, which may be used to guide humans step by step through the physical task execution process by providing them with *cognitive assistance*. Technically this could be easily realized by equipping humans with an augmented reality (AR) headset (e.g. HoloLens 2 with WiFi connectivity) that receives work-order information in real-time from its nearest AI-enhanced MEC server. Recall that in Section 2.2, we elaborated on the importance of shifting the research focus from AI to intelligence amplification (IA) by using information technology to enhance human decisions. Note, however, that IA becomes difficult in dynamic task environments of increased uncertainty and real-word situations of great complexity.

6.5.2 HITL Hybrid-Augmented Intelligence

Many problems that humans encounter tend to be highly uncertain, complex, and open-ended. Human interaction and participation must be introduced to solve such problems, giving rise to the concept of *HITL hybrid-augmented intelligence* for advanced human-machine collaboration (Zheng et al., 2017). HITL hybrid-augmented intelligence is defined as an intelligent model that calls for human input and enables solving problems that may not be easily addressed by machine learning. In general, machine learning is inferior to the human brain in understanding unstructured real-world environments and processing incomplete information and complex spatiotemporal correlation tasks. Hence, machines cannot carry out all the tasks in human society on their own. Instead, AI and human intelligence are better viewed as highly complementary.

According to Zheng et al. (2017), the Internet provides an immense innovation space for HITL hybrid-augmented intelligence. Specifically, cloud robotics and AR are among the fastest growing commercial applications for enhancing the intelligence of an individual in multi-robot collaborative systems. One of the key research avenues of HITL hybrid-augmented intelligence is the development of methods that allow machines to learn from not only massive training samples but also human knowledge in order to execute highly intelligent tasks via shared intelligence among different robots and humans.

6.5.3   Decentralized Self-Organizing Cooperative (DSOC)

A very interesting example of converting human and machine intelligence into a new form of self-organizing artificial general intelligence (AGI) across the Internet is the so-called *SingularityNET* (https://singularitynet.io). One can think of Singu-larityNet as a *decentralized self-organizing cooperative (DSOC)*, a concept similar to DAO. DSOC is essentially a distributed computing architecture for making new kinds of smart contracts. Entities executing these smart contracts are referred to as agents, which can run in the cloud, on phones, robots, or other embedded devices. Services are offered to any customer via APIs enabled by smart contracts and may require a combination of actions by multiple agents using their collective intelligence. In general, there may be multiple agents that can accomplish a given task request in different ways and to different degrees. Each task request to the network requires a unique combination of agents, thus forming a so-called *offer network* of mutual dependency, where agents make offers to each other to exchange services via offer-request pairs. Whenever someone wants an agent to perform services, a smart contract is signed for this specific task. Toward this end, DSOC aims at leveraging contributions from the broadest possible variety of agents by means of superior discovery mechanisms for finding useful agents and *nudging* them to become contributors.

6.5.4   Nudge Contract: Nudging via Smart Contract

Extending on DSOC and the JC introduced in Section 6.1, we develop a *nudge contract* for enhancing the human capabilities of unskilled crowd members of the DAO. According to Richard H. Thaler, the 2017 Nobel Laureate in Economics, a nudge is defined as any aspect of a choice architecture that changes people's behavior in a predictable way, while not ruling out any options nor significantly changing their economic incentives. Deployed appropriately, nudges can *steer people*,

---

Algorithm 7 Nudge Contract

Input: Set $U = \{h_1, h_2, \ldots, h_n\}$ of n DAO members, capability vector $\mathbf{C} = [c_1, c_2, \ldots, c_n]$, distance vector $\mathbf{D} = [d_1, d_2, \ldots, d_n]$, interrupted task $\mathbf{T}$, required number D of actions to execute the interrupted task, interrupted robot r_0, capability requirement c_0 of the interrupted task

1: Decompose the given interrupted task $\mathbf{T}$ into N_{sub} subtasks
2: **for** $i = 1$ to n **do**
3:    **if** $c_i \geq c_0$ **then**
4:      $S \leftarrow h_i$
5:    **end if**
6: **end for**
7: $h^* \leftarrow \arg\min_{d_i}\{S\}$
8: Create a secure blockchain transaction between h^* and interrupted robot r_0
9: Send the learning instructions from h^* to r_0 through the established transaction
10: Use the multi-arm bandit selection strategy in McGuire et al. [2018] to help the robot learn the given set of subtasks
11: **if** all N_{sub} subtasks are learned successfully **then**
12:    learning process is successfully accomplished
13:    r_0 can execute the interrupted task $\mathbf{T}$ with the capability of h^*
14: **else**
15:    Learning process is failed
16:    DAO member h^* traverses to the interruption point to execute the task $\mathbf{T}$
17: **end if**
18: Reward the skilled DAO member h^* via blockchain smart contract

---

as opposed to steer objects – real or virtual – as done in the conventional Tactile Internet, to make better choices and positively influence the behavior of crowds of all types.

Our nudge contract aims at completing interrupted physical tasks by learning from a skilled DAO member with the objective of minimizing the learning loss, which denotes the difference between the achievable and optimum task execution times (McGuire et al., 2018). The ability to learn a given subtask is characterized by the subtask learning probability. The learning process is accomplished if each subtask is learned successfully from a skilled DAO member, who in turn is rewarded via a smart contract (see Algorithm 7 for details). Figure 6.5 shows the performance of our nudge contract for 50 DAO crowd members, whose ratio $\frac{f_{human}}{f_{robot}}$ is randomly chosen from $\{1/3, 3/3, 5/3\}$. We observe that for a given subtask learning probability, decreasing the number N_{sub} of subtasks helps reduce the learning loss, thus indicating the importance of a proper task decomposition method.

Figure 6.5 Learning loss (in seconds) vs. subtask learning probability for different number N_{sub} of subtasks and traverse time $t_{traverse}$.

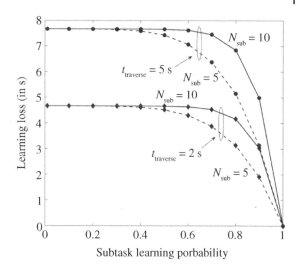

6.6 Conclusions

In this chapter, we explored how Ethereum blockchain technologies, in particular the DAO, may be leveraged to decentralize the Tactile Internet, which enables unprecedented mobile applications for remotely steering real or virtual objects/processes in perceived real-time and represents a promising example of future techno-social systems. We showed that a higher level of decentralization of AI-enhanced MEC reduces the average computational task completion time of up to 89.5% by setting the computation offloading probability to 0.7. Further, we observed that crowdsourcing of human assistance is beneficial in decreasing the average completion time of physical tasks for medium to high feedback misforecasting probabilities, provided the human offers equal or even superior operational capabilities, i.e. $\frac{f_{human}}{f_{robot}} \geq 1$. Toward this end, our proposed nudge contract tries to successfully accomplish tasks via shared intelligence among failing robots and skilled humans.

7

XR in the 6G Post-Smartphone Era

7.1 Introduction

At the 2015 World Economic Forum, Eric Schmidt famously stated that "the Internet will disappear" given that there will be so many things that we are wearing and interacting with that we won't even sense the Internet, though it will be part of our presence all the time. Although this first might sound a bit surprising, it is actually what profound technologies do in general. In "The Computer for the 21st Century," Mark Weiser argued that the most profound technologies are those that disappear. They weave themselves into the fabric of everyday life until they are indistinguishable from it Weiser (1999).

An interesting recent approach to make the Internet disappear is the so-called Naked world vision that aims at paving the way to an *Internet of No Things* by offering all kinds of human-intended services without owning or carrying any type of computing or storage devices (Ahmad et al., 2018). The term Internet of No Things was coined by Demos Helsinki founder RoopeMokka in 2015. The term nicely resonates with Eric Schmidt's aforementioned statement. The Naked world envisions Internet services to appear from the surrounding environment when needed and disappear when not needed. The transition from the current gadgets-based Internet to the Internet of No Things is divided into three phases that starts from *bearables* (e.g. smartphone), moves toward *wearables* (e.g. Google and Levi's smart jacket or Amazon's recently launched voice-controlled Echo Loop ring, glasses, and earbuds), and then finally progresses to the last phase of so-called *nearables*. Nearables denote nearby surroundings or environments with embedded computing/storage technologies and service provisioning mechanisms that are intelligent enough to learn and react according to user context and history in order to provide user-intended services. According to Ahmad et al. (2018), their successful deployment is challenging not only from a technological point of view but also from a

Toward 6G: A New Era of Convergence, First Edition. Amin Ebrahimzadeh and Martin Maier.
© 2021 The Institute of Electrical and Electronics Engineers, Inc.
Published 2021 by John Wiley & Sons, Inc.

business and social mindset perspective due to the required user acceptability and trust.

Some of the most interesting 5G applications – most notably, virtual reality (VR) and Tactile Internet – seem to evolve in the same direction. To see this, note that according to Bastug (2017), VR systems will undergo three evolutionary stages, similar to the aforementioned Internet of No Things. The first evolutionary stage includes current VR systems that require a wired connection to a PC or portable device because current 4G or even pre-5G wireless systems cannot cope with the massive amount of bandwidth and latency requirements of VR. The PC or portable device in turn is connected to the central cloud and the Internet via backhaul links. At the second evolutionary stage, VR devices are wirelessly connected to a fog/edge server located at the base station (BS) for local computation and caching. The third and final evolutionary stage envisions ideal (fully interconnected) VR systems, where no distinction between real and virtual worlds are made in human perception. In addition, according to Bastug (2017), the growing number of drones, robots, and self-driving vehicles will take cameras to places humans could never imagine reaching. Similarly, the Tactile Internet, specified within the IEEE P1918.1 standards working group, allows for the tactile steering and control of not only virtual but also real objects (e.g. teleoperated robots) as well as processes. Thus, the Tactile Internet may be viewed as an extension of immersive VR from a virtual to a physical environment. Recall that in Chapter 2, we showed that the human-centric design approach of the Tactile Internet helps extend the capabilities of humans through the Internet by supporting them in the coordination of their physical and digital co-activities with robots and software agents by means of (artificial intelligence) AI-enhanced multi-access edge computing (MEC).

The above discussion shows that future fully interconnected VR systems and the Tactile Internet seem to evolve toward common design goals. Most notably, the boundary between virtual (i.e. online) and physical (i.e. offline) worlds is to become increasingly imperceptible while both digital and physical capabilities of humans are to be extended via edge computing variants, ideally with embedded AI capabilities. According to the inaugural report "Artificial Intelligence and Life in 2030" of Stanford University's recently launched One Hundred Year Study on artificial intelligence (AI100), an increasing focus on developing systems that are human-aware is expected over the next 10–15 years.

In this chapter, we elaborate on how the Internet of No Things with its underlying human-intended services may serve as a useful stepping stone toward realizing the far-reaching vision of future 6G networks, ushering in the *6G post-smartphone era*. After briefly reviewing the 6G vision, we explain the reality–virtuality continuum in more detail and introduce the so-called *Multiverse* for the design of advanced *extended reality (XR)* experiences, ranging from conventional VR to more sophisticated cross-reality environments known as third spaces. We then elaborate on the recently emerging *invisible-to-visible (I2V)* technology concept, which we use together with other key enabling network technologies to tie both online and offline worlds closer together in an Internet of No Things and make it "see the invisible" through the awareness of nonlocal events in space and time.

7.2   6G Vision: Putting (Internet of No) Things in Perspective

The authors of Letaief et al. (2019) provided a roadmap to 6G, which envisions that, in contrast to previous generations, 6G will be transformative and will revolutionize the wireless evolution from "connected things" to "connected intelligence." According to Strinati et al. (2019), 6G will play a significant role in advancing Nikola Tesla's prophecy that "when wireless is perfectly applied, the whole Earth will be converted into a huge brain." Toward this end, the authors of Strinati et al. (2019) argue that 6G will provide an information and communication technology (ICT) infrastructure that enables end-users to perceive themselves as surrounded by a huge artificial brain offering virtually zero-latency services, unlimited storage, and immense cognitive capabilities. In 6G, there is also a strong notion that the nature of mobile terminals will change, whereby smart cars and intelligent mobile robots are anticipated to play a more important role (Zong et al., 2019).

6G is anticipated to allow for the inclusion of additional human sensory information. The ITU Telecommunication standardization sector (ITU-T) Focus Group Technologies for Network 2030 (FG NET-2030) was established in July 2018 to study and advance the capabilities of the networks for the year 2030 and beyond. Among others, FG NET-2030 envisions user experiences to go from well-explored audio–visual communications to the delivery of all five human senses as well as *other senses* in line with the IEEE Digital Senses Initiative. In David and Berndt (2018), the authors advocate that 6G should embrace a new mode of thinking from the get-go by including social awareness and understanding the social impact of advanced technologies. They argue that deepened personalization of 6G services that could *predict future events* for the user and provide good advice would certainly be appreciated.

Finally, the authors of Saad et al. (2020) observed that the ongoing deployment of 5G cellular systems is exposing their inherent limitations compared to the original premise of 5G as an enabler for the Internet of Everything (IoE). They argue that 6G should not only explore more spectrum at high-frequency bands but, more importantly, converge driving technological trends, thereby ushering in the 6G post-smartphone era. Their bold, forward-looking research agenda intends to serve as a basis for stimulating more out-of-the-box research that will drive the

6G revolution. Specifically, they claim that there will be the following four driving applications behind 6G: (i) *multisensory XR* applications, (ii) *connected robotics* and autonomous systems, (iii) wireless brain-computer interaction (a subclass of *human–machine interaction*), and (iv) *blockchain* and distributed ledger technologies. Among other 6G driving trends and enabling technologies, they emphasize the importance of haptic and empathic communications, edge AI, the emergence of smart surfaces/environments and new human-centric service classes, as well as the end of the smartphone era, given that smart wearables are increasingly replacing the functionalities of smartphones. They also expect that research on the *quantum realm* will intersect with 6G toward its end of standardization.

The Internet of No Things with its underlying human-intended services and nonlocal extension of human "sixth-sense" experiences in both space and time may serve as a useful stepping stone toward realizing the far-reaching 6G vision above, as explained in technically greater detail in the remainder of the chapter.

7.3 Extended Reality (XR): Unleashing Its Full Potential

In this section, we further elaborate on the recently emerging term XR and how its full potential can be unleashed.

7.3.1 The Reality–Virtuality Continuum

According to Qualcomm, XR will be the next-generation mobile computing platform that brings the different forms of reality together in order to realize the entire reality–virtuality continuum of Figure 7.1 for the extension of human experiences, including the support of human–machine interaction. In fact, according to a recent ABI Research and Qualcomm study, some of the most exciting XR use cases include remotely controlled devices and the Tactile Internet (ABI Research and Qualcomm, 2017).

The reality–virtuality continuum ranges from pure reality (offline) to pure virtuality (online), as created by VR. Both reality and virtuality may be augmented, leading to augmented reality (AR) on one side of the continuum and augmented virtuality (AV) on the other. AR enables the live view of a physical, real-world environment, whose elements are augmented by computer-generated perceptual information, ideally across multiple sensory modalities. In doing so, AR alters one's perception of the real-world environment, as opposed to VR, which replaces the real-world environment with a simulated one. Conversely, AV occurs in a virtual environment, where a real object is inserted into a computer-generated environment.

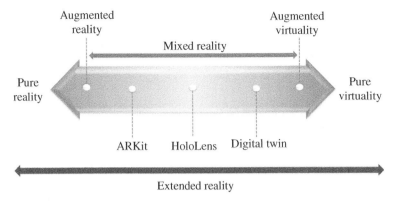

Figure 7.1 The reality–virtuality continuum, ranging from pure reality (offline) to pure virtuality (online).

A good AR example is Shopify's ARKit, which allows smartphones to place 3D models of physical items and see how they would look in real life. To do so, AR overlays virtual objects/images/information on top of a real-world environment. An illustrative AV example is an aircraft maintenance engineer, who is able to visualize a real-time model, often referred to as digital twin, of an engine that may be thousands of kilometers away. The term mixed reality (MR) includes AR, AV, and mixed configurations thereof, blending representations of virtual and real-world elements together in a single user interface. MR helps bridge the gap between real and virtual environments, whereby the difference between AR and AV reduces to where the user interaction takes place. If the interaction happens in the real world, it is considered AR. By contrast, if the interaction occurs in a virtual space, it is considered AV. The flagship MR device is Microsoft's HoloLens 2. The areas where most industries apply XR is in remote guidance systems for performing complex tasks such as maintenance and assembly (Fast-Berglund et al., 2018).

7.3.2   The Multiverse: An Architecture of Advanced XR Experiences

Apart from VR/AR/MR, future XR technologies may realize novel, unprecedented types of reality. Thus, X may be rather viewed as a placeholder for future yet unforeseen developments on the digital frontier. An interesting attempt to charter the unknown territory is the so-called *Multiverse*, which may serve as an architecture of advanced XR experiences (Pine and Korn, 2011). As shown in Figure 7.2, the Multiverse consists of the following architectural components:

- *Dimensions*: There are the three well-known physical dimensions – Space, Time, and Matter – that constitute our physical reality.

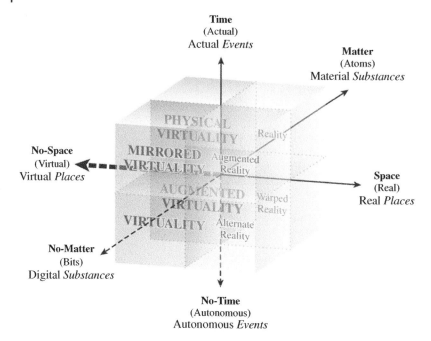

Figure 7.2 The Multiverse as an architecture of advanced XR experiences: three dimensions, six variables, and eight realms. Source: (Pine and Korn, 2011). © Berrett-Koehler Publishers.

- *Variables*: In addition, there are three nonphysical dimensions – referred to as *No-Space, No-Time*, and *No-Matter* – that make up the virtual world. Unlike their physical counterparts, these three digital dimensions are not subject to the constraints imposed by physical space, time, and matter. Thus, in total there are six variables that can be exploited for the design of advanced XR experiences.
- *Realms*: Given that there are three (3) pairs of variables, each with two (2) opposite physical/digital dimensions, we have a total of $2^3 = 8$ possible realms. Each realm creates a different type of reality, ranging from conventional AR/AV to more sophisticated types of reality, e.g. mirrored virtuality, warped reality, and alternate reality. *Mirrored virtuality* absorbs the real world into the virtual and creates a virtual expression of reality that unfolds as it actually happens, providing a particular bird's eye view. *Warped reality* plays with time in any way possible by taking an experience firmly grounded in reality and shifting it from actual to autonomous time. *Alternate reality*, on the other hand, creates an alternative view of the real world by constructing a digital experience

and superimposing it onto a real place. Unlike AR, however, alternate reality manipulates time and allows looking to the future freed from the bonds of actual time.

According to Pine and Korn (2011), the Multiverse with its different variables and realms offers a powerful experience design canvas to uncover hidden XR opportunities by fusing the real and the virtual, thereby creating *cross-reality environments* or so-called third spaces. Third spaces are created whenever one transverses the boundary between realms within any given experience. It is worthwhile to mention that, in Weiser (1999), Mark Weiser seems to had something similar in mind when describing what he initially called *embodied virtuality*, which is now more widely referred to as ubiquitous computing.

In the subsequent section, we explore how the above concepts (No-Space, No-Time, No-Matter, realms, cross-reality environments) can be used to tie both online and offline worlds closer together in an Internet of No Things and make it "see the invisible."

7.4 Internet of No Things: Invisible-to-Visible (I2V) Technologies

Recall from Section 7.1 that future fully interconnected VR systems will leverage on the growing number of drones, robots, and self-driving vehicles. A very interesting example of future connected-car technologies that merges real and virtual worlds to help drivers "see the invisible" is Nissan's recently unveiled *I2V* technology concept (Nissan Newsroom, 2019). I2V creates a three-dimensional immersion connected-car experience that is tailored to the driver's interests by changing how cars are driven and integrated into society. More specifically, by merging information from sensors outside and inside the vehicle with data from the cloud, I2V enables the driver and passengers not only to track the vehicle's immediate surroundings but also to anticipate what's ahead, e.g. what's behind a building or around the corner. Although the initial I2V proof-of-concept demonstrator used AR headsets (i.e. wearables), Nissan envisions to turn the windshield of future self-driving cars into a portal to the virtual world, thus finally evolving from wearables to nearables, as discussed in Section 7.1 in the context of the Internet of No Things.

I2V is powered by Nissan's *omnisensing* technology, a platform originally developed by the video gaming company Unity Technologies, which acts as a hub gathering real-time data from the traffic environment and from the vehicle's surroundings and interior to anticipate when people inside the vehicle may need assistance. The technology maps a 360° virtual space and gives guidance in an interactive, human-like way, such as through avatars that appear inside the car. It can also connect passengers to people in the so-called *Metaverse* virtual world that is shared with other users. In doing so, people may appear inside the car as AR avatars to provide assistance or company. For instance, when visiting a new place, I2V can search within the Metaverse for a knowledgeable local guide. The information provided by the guide may be stored in the cloud such that others visiting the same area can access it or may be used by the onboard AI system for a more efficient drive through local areas. Alternatively, the driver may book a professional driver from the Metaverse, who appears as a virtual chase car in the driver's field of view to show the best way and improve driving skills, just like in a video game.

Clearly, I2V opens up endless opportunities by tapping into the virtual world. In fact, the IEEE P1918.1 standard, briefly mentioned in Section 7.1, highlights several key use cases of the Tactile Internet, including not only the automative control of connected/autonomous driving via virtual avatars but also the remote control of physical robots. According to Haddadin et al. (2019), the vastly progressing smart wearables such as exoskeletons and VR/AR devices effectively create real-world avatars, i.e. tactile robots connected with human operators via smart wearables, as a central *physical embodiment* of the Tactile Internet. More specifically, the authors of Haddadin et al. (2019) argue that the Tactile Internet creates the new paradigm of an immersive coexistence between humans and robots in order to achieve tight physical human–robot interaction (pHRI) and entanglement between man and machine in future locally connected *human–avatar/robot collectives*. Assistive exoskeletons are thereby envisaged to become an important element of the Tactile Internet in that they extend user capabilities or supplement/replace some form of function loss, e.g. lifting heavy objects or rehabilitation systems for people with spinal cord injury. In addition, many studies have shown that the physical presence of robots benefited a variety of social interaction elements such as persuasion, likeability, and trustworthiness. Thus, leveraging these beneficial characteristics of social robots represents a promising solution toward addressing the user acceptability and trust issues of nearables mentioned in Section 7.1.

In the following, we build on the I2V technology concept and explore how emerging multisensory XR technologies in conjunction with AI-enhanced MEC, intelligent mobile robots, and blockchain technologies may be combined to usher in the Internet of No Things as an important stepping stone toward realizing the 6G vision outlined in Section 7.2.

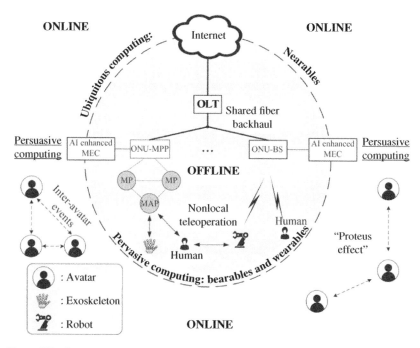

Figure 7.3 Extrasensory perception network (ESPN) architecture integrating ubiquitous, pervasive, and persuasive computing.

7.4.1 Extrasensory Perception Network (ESPN)

Let our point of departure be Joseph A. Paradiso's pioneering work on extrasensory perception (ESP) in an Internet of Things (IoT) context at MIT Media Lab (Dublon et al., 2014). In a sensor-driven world, network-connected sensors embedded in anything function as extensions of the human nervous system and enable us to enter the long-predicted era of ubiquitous computing, as envisioned by Mark Weiser more than a quarter of century ago (see Section 7.3). In Dublon et al. (2014), the authors showed that network-connected sensors and computers make it possible to virtually travel to distant environments and "be" there in real time. Interestingly, the authors concluded that future technologies will fold into our surroundings that help us to get our noses off the smartphone screens and back into our environments, thus making us more (rather than less) present in the world around us. Clearly, this human-centric outlook on future technologies may materialize in the 6G post-smartphone era.

Recall from Section 7.3.1 that XR will be the next-generation mobile computing platform for the extension of human experiences, including the support of human–machine interaction. Figure 7.3 depicts the architecture of our proposed *extrasensory perception network (ESPN)*, which integrates the following three evolutionary stages of mobile computing: (i) ubiquitous, (ii) pervasive, and (iii) persuasive computing. Ubiquitous computing is embedded in the things surrounding us (i.e. nearables), while pervasive computing involves our bearables and wearables. Persuasive computing aims at changing the behavior of users through social influence. An interesting phenomenon for changing behavior in an online virtual environment is the so-called "Proteus effect," where the behavior of individuals is shaped by the characteristics and traits of their virtual avatars, especially through interaction during inter-avatar events. We will exploit AI-enhanced MEC to realize persuasive computing, as described in more detail shortly.

We have seen in Section 7.3.1 that some of the most exciting XR use cases include remotely controlled devices and the Tactile Internet (ABI Research and Qualcomm, 2017). Recall from Chapter 2 that we studied the Tactile Internet as one of the most interesting 5G low-latency applications enabling novel immersive experiences by means of haptic communications (Maier and Ebrahimzadeh, 2019). Recall also that the emerging Tactile Internet will remain a prominent application enabled by future 6G mobile networks. The underlying physical network infrastructure, which is illustrated in Figure 7.3, consisted of a fiber backhaul shared by wireless local area network (WLAN) mesh portal points (MPPs) and cellular BSs that are collocated with optical network units (ONUs), which in turn are connected to the central optical line terminal (OLT) of the fiber backhaul. Based on real-world haptic traces, we studied the use case of *nonlocal teleoperation* between an human operator (HO) and teleoperator robot (TOR), which are both physical, i.e. offline, entities (see also Figure 7.3). We showed that AI-enhanced MEC helps decouple haptic feedback from the impact of extensive propagation delays by forecasting delayed or lost haptic feedback samples. This enables humans to perceive remote task environments in real-time at a 1-ms granularity.

7.4.2 Nonlocal Awareness of Space and Time: Mimicking the Quantum Realm

As an illustrative example of advanced XR experiences, we study the delivery of extrasensory human perceptions, i.e. senses other than the five human senses, as envisioned by the IEEE Digital Senses Initiative and ITU-T FG NET-2030 (see Section 7.2).

It is interesting to note that the term ESP actually refers to a widely known phenomenon that allows humans to have nonlocal experiences in space and time. According to Wikipedia, ESP is also called *sixth sense*, which includes claimed reception of information not gained through the recognized five physical senses, but sensed with the mind. There exist different types of ESP, including clairvoyance (i.e. viewing things or events at remote locations) and precognition (i.e. viewing future events before they happen). While clairvoyance may be viewed as the ability to perceive the hidden present, precognition is a forecast (not prophecy) of events to come about in the future unless one does something to change them based on the perceived information. In contemporary physics, there exists the so-called "principle of nonlocality," also referred to as quantum-interconnectedness of all things by quantum physicists such as David Bohm, which transcends spatial and temporal barriers (Bohm, 2002). Nonlocality occurs due to the phenomenon of entanglement, where a pair of particles have complementary properties when measured, and might be the cause of ESP.

Phenomena such as entanglement also play an important role in the nascent *Quantum Internet* Cacciapuoti et al. (2020). The Quantum Internet consists of both classical and quantum links interconnecting remote quantum devices. With respect to quantum communication resources, it seems attractive to utilize existing optical fiber networks. However, it is still an open problem to determine whether it is feasible to utilize a single link, e.g. a single optical fiber, for both quantum and classical communications, such that existing network infrastructures can be exploited without the need for additional infrastructures. Hence, from a communication engineering perspective, the design of the Quantum Internet is not an easy task at all since it is governed by the laws of quantum phenomena with no counterpart in classical networks, which impose serious constraints on the network design. A key strategy for transmitting information in the Quantum Internet is *teleportation*. Quantum teleportation provides an invaluable strategy for transmitting so-called quantum bits (qubits) without either the physical transfer of the particle storing the qubit or the violation of the quantum mechanics principles.

According to Cacciapuoti et al. (2020), the Quantum Internet is probably still a concept far from real-world implementation. In addition, the quantum teleportation process, which represents the core communication functionality of the Quantum Internet, is gravely affected by a number of quantum imperfections that arise during the quantum teleportation process from a communication engineering perspective (Cacciapuoti et al., 2020a). Despite the fact that the Quantum Internet might pave the way for the Internet of the future, there is a substantial amount of frontier-research required for tackling the challenges and open problems associated with it. By contrast, with the advent of advanced XR technologies it might be easier to mimic the Quantum Internet instead of actually building it, as explained in more detail next.[1]

Note that despite reports based on anecdotal evidence, there has been no convincing scientific evidence that ESP exists after more than a century of

---

1 Still unsettled is the discussion about whether the brain is a natural quantum computer or not. Nevertheless, it is worthwhile to mention that there are many theories that in some way relate the brain to quantum physics, where quantum effects play some kind of role in the brain.

research. However, instead of rejecting ESP as pseudoscience, in this chapter we argue that with the emergence of XR it might become possible to disrupt the old impossible/possible boundary and mimic the quantum realm. Toward this end, we are going to design the following two advanced XR experiences that transverse the boundary between the aforementioned Multiverse realms in order to realize awareness of nonlocal events in space and time.

7.4.2.1 Precognition

To achieve precognition, we extend our aforementioned AI-enhanced MEC based haptic feedback sample forecasting scheme in Chapter 2 for realizing persuasive computing. Recall from above that in nonlocal teleoperation the HO and TOR are physical entities, i.e. both reside in the realm *reality* characterized by Space, Time, and Matter. In addition, we let the HO have access to the realm *AV* (i.e. No-Space, Time, No-Matter) by observing a digital twin of the remote TOR via a wearable head-mounted display.

Our AI-enhanced MEC forecasting scheme was trained by using haptic traces obtained from application-specific teleoperation experiments. It was shown in Chapter 2 that a high forecasting accuracy (mean squared error below 1‰) can be achieved in the considered scenarios. In general, however, the training may become irrelevant in changing or unstructured real-world environments, resulting in a decreased forecasting accuracy. How can the HO know when or even before this happens and be persuaded to make an informed decision?

To quantify the decreasing effectiveness, our AI-enhanced MEC computes the metric regret, which measures the future regret the HO will have after *blindly* relying on a presumably intact haptic feeback sample forecasting scheme. We define regret as the difference between the achievable and the optimum physical task execution times of the TOR. Note that the metric regret is used to influence the HO's decision to abort the teleoperation before unintended consequences might occur. It is displayed in his head-mounted wearable to "make him see" the AI becoming less trustworthy. In Section 7.5, we will highlight some illustrative results.

7.4.2.2 Eternalism

Next, we consider also the transition from the Time to No-Time dimension of the Multiverse in Figure 7.2. In physics, the two most important theories on the nature of time have been *presentism* and *eternalism*. Presentism states that only the present is real. By contrast, eternalism states that the past and future are as equally real as the present. Under eternalism, "now" is to time as "where" is to space, whereby time is a dimension much like space, one in which the past and future are as real as locations north and south (i.e. unlike presentism, eternalism thus lends itself to time travel). Today, most physicists view eternalism as the order of time (Buonomano, 2017).

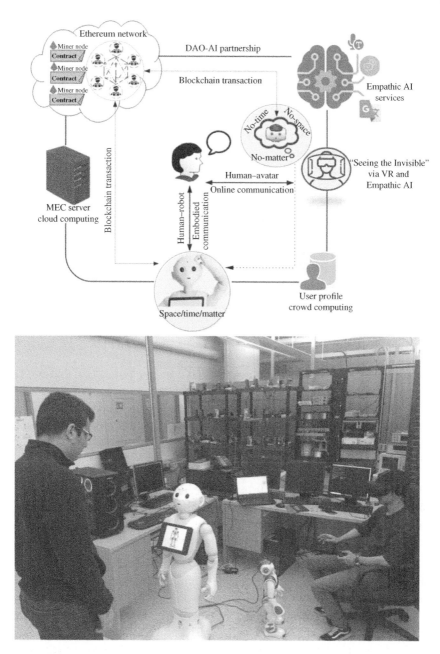

Figure 7.4 Experimental set-up for demonstrating eternalism in locally connected human–avatar/robot collectives.

Figure 7.4 illustrates our experimental set-up for demonstrating eternalism in locally connected human–avatar/robot collectives. The human engages in embodied communication with Pepper, SoftBank Mobile's most advanced humanoid robot, via voice, gesture, and Pepper's built-in Android tablet. We use an Oculus Rift VR headset to let the human also access the virtual avatar of the robot. A user profile is maintained to record each human–robot interaction. In addition, we exploit IBM Watson's empathic AI services, most notably, IBM's tone analyzer for detecting emotions in written text exchanged during human–robot/avatar online communication. As blockchain of choice we deploy Ethereum to interconnect all human operators, real and virtual robots, and empathic AI services in a decentralized autonomous organization (DAO), one of Ethereum's salient features, to share skills and help solve complex problems faced during human–robot–avatar interactions. MEC-based cloud computing is used for offloading compute-intensive blockchain transactions, e.g. mining, from resource-limited robots. In the subsequent section, we highlight a use case of exploiting VR and empathic AI to *make emotions visible* and nudge the human toward experiencing eternalism.

7.5 Results

Let us consider an HO–TOR pair carrying out a given physical task that can be decomposed into 100 operations. To achieve the optimum task execution time, the AI-enhanced MEC forecasting scheme outsources certain operations to another crowdsourced HO, who is located 20 seconds away from the physical task point. Let f_H and f_R denote the capability (given in number of operations per second) of the HO and TOR to execute the physical task, respectively. The HO decides to abort teleoperation, when he observes the digital twin starting to produce failures and the ratio of misforecast samples to total number of received haptic feedback samples exceeds a certain threshold S_H. Subsequently, the crowdsourced HO traverses to the physical task point to finalize all remaining operations.

Figure 7.5 depicts the regret vs. misforecast sample rate λ_f for different ratio $\frac{f_H}{f_R}$ and S_H. It highlights the beneficial role of crowdsourcing a capable assistant HO with increased $\frac{f_H}{f_R}$ in compensating for an unreliable AI and completing the physical task failure-free.

Note that the above digital twin is synchronized with the remote TOR, both operating in the actual Time dimension of Figure 7.2. Next, we also tap into the No-Time dimension of VR environments during the following time travel experiment from reality to virtuality. The experiment lasted 15 minutes and was repeated five times, each time involving a different student. In the initial reality

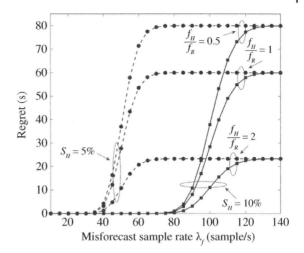

Figure 7.5 Regret (given in seconds) vs. misforecast sample rate λ_f for different ratios of human and robot capabilities $\frac{f_H}{f_R} \in \{0.5, 1, 2\}$ with human decision threshold $S_H \in \{5\%, 10\%\}$.

part, the student first engages with Pepper for an interactive audio-visual tour of INRS (the students' university). Subsequently, the student is given the opportunity to ask Pepper any arbitrary question about INRS, whereby Pepper's responses are provided by a remote human operator via speech-to-text and text-to-speech conversion. Next, Pepper invites the student to continue the experiment in the virtuality part, where the student can virtually walk through INRS guided by an avatar acting as an omniscient oracle. The oracle relies on a remote human operator, who is able to monitor the student's detected emotions in real-time. By leveraging on the "Proteus effect" experienced in inter-avatar events (see Figure 7.3), the oracle gives advice to the student on how to gradually reach a desirable future situation at INRS, which is characterized by higher levels of confidence and emotional engagement.

Figure 7.6 shows the average empathic AI score of the four positive emotions detected by IBM Watson's tone analyzer during the various human–robot/avatar speech-to-text-to-speech exchanges of the experiment. It clearly illustrates that the students become increasingly more confident and emotionally less tentative after transiting from reality to virtuality, thus confirming the beneficial impact of the Proteus effect.

7.6 Conclusions

Our proposed ESPN architecture integrated ubiquitous, pervasive, and persuasive computing to enable the delivery of so-called extrasensory *sixth-sense* human perceptions via advanced XR experiences in a future Internet of No Things, which will

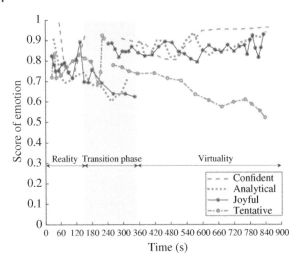

Figure 7.6 Average empathic AI score of four different positive emotions experimentally detected during time travel from reality to virtuality.

be increasingly based on wearables (e.g. VR headsets) and nearables (e.g. intelligent mobile robots) in an anticipated 6G post-smartphone era. We exploited the so-called Multiverse concept to design cross-reality environments that help fuse the real and the virtual in networked human–avatar/robot collectives. By means of simulation and experiment, we studied two illustrative cross-reality use cases to make humans see AI becoming less trustworthy and to exploit empathic AI services making human emotions visible.

Appendix A

Proof of Lemmas

A.1 Proof of Lemma 3.1

Proof: Using Eq. (3.1) along with the velocity profile shown in Fig. 3.2, the energy consumption E_{trav} of the MR to traverse a given distance Δd is calculated as follows:

$$E_{trav} = E(\omega_d) = \underbrace{\int_0^{T_{trav}} c_1 a^2(t).dt}_{E_1} + \underbrace{\int_0^{T_{trav}} c_2 v^2(t).dt}_{E_2}$$

$$+ \underbrace{\int_0^{T_{trav}} c_3 v(t).dt}_{E_3} + \underbrace{\int_0^{T_{trav}} c_4.dt}_{E_4}. \tag{A.1}$$

Note that for the considered velocity profile $v(t)$ in Fig. 3.2, the contributions of both fifth and sixth terms in Eq. (3.1) are equal to zero, as $\int_0^{T_{trav}} a(t).dt = 0$ and $\int_0^{T_{trav}} v(t)a(t).dt = 0$. We obtain E_1, E_2, E_3, and E_4 as follows:

$$E_1 = c_1 \left(\int_0^{t_1} a_{acc}^2.dt + 0 + \int_{t_2}^{t_3} a_{dec}^2.dt \right)$$

$$= c_1 \left(\int_0^{T_{acc}} \left(\frac{v_{max}}{T_{acc}} \right)^2.dt + \int_{T_{acc}+T_{cst}}^{T_{trav}} (-\frac{v_{max}}{T_{dec}})^2.dt \right)$$

$$\overset{Eq.(3.4)}{=} 2c_1(1-\omega_d)\frac{\Delta d}{v_{max}} \left(\frac{v_{max}^2}{(1-\omega_d)\Delta d} \right)^2 = \frac{2c_1 v_{max}^3}{(1-\omega_d)\Delta d}, \tag{A.2}$$

$$E_2 = c_2 \left(\int_0^{t_1} \left(\frac{v_{max}}{T_{acc}} t \right)^2.dt + \int_{t_1}^{t_2} (v_{max})^2.dt + \int_{t_2}^{t_3} \left(-\frac{v_{max}}{T_{acc}}(t - T_{trav}) \right)^2.dt \right)$$

Toward 6G: A New Era of Convergence, First Edition. Amin Ebrahimzadeh and Martin Maier.
© 2021 The Institute of Electrical and Electronics Engineers, Inc.
Published 2021 by John Wiley & Sons, Inc.

$$\overset{\text{Eq.(3.4)}}{=} c_2 \left(\frac{2}{3}(1 - \omega_d)v_{max}\Delta d + \omega_d v_{max}\Delta d \right), \tag{A.3}$$

$$E_3 = c_3 \int_0^{T_{trav}} v(t).dt \overset{\text{Eq.(3.3)}}{=} c_3 \Delta d, \tag{A.4}$$

$$E_4 = c_4 \int_0^{T_{trav}} 1.dt \overset{\text{Eq.(3.5)}}{=} c_4(2 - \omega_d)\frac{\Delta d}{v_{max}}. \tag{A.5}$$

Substituting Eqs. (A.2)-(A.5) into Eq. (A.1), completes the proof. □

A.2 Proof of Lemma 3.2

Proof: Given $\frac{\partial^2 E(\omega_d)}{\partial \omega_d^2} > 0$ for $\omega_d \in (0,1)$, in order for $E(\omega_d)$ to have a local minimum in interval $(0,1)$, $\frac{\partial E(\omega_d)}{\partial \omega_d}$ has to be zero. Therefore, we have

$$\frac{\partial E(\omega_d)}{\partial \omega_d} = \frac{v_{max}c_2\Delta d}{3} + \frac{2c_1 v_{max}^3}{\Delta d(1 - \omega_d)^2} - \frac{c_4\Delta d}{v_{max}} = 0, \tag{A.6}$$

which gives $\hat{\omega}_d$ as follows:

$$\hat{\omega}_d = 1 - \sqrt{\underbrace{\frac{\frac{6c_1 v_{max}^4}{\Delta d^2}}{3c_4 - v_{max}^2 c_2}}_{M'}}. \tag{A.7}$$

We note that $\hat{\omega}_d$ has to lie in interval $(0,1)$, thus implying that

$$0 < M' < 1. \tag{A.8}$$

The left-hand inequality holds for $v_{max} < v_1'$, where $v_1' = \sqrt{\frac{3c_4}{c_2}}$. Whereas the right-hand inequality translates into

$$\underbrace{\frac{6c_1}{\Delta d}v_{max}^4 + c_2 v_{max}^2 - 3c_4}_{Q_E(v_{max})} < 0. \tag{A.9}$$

To evaluate the range of v_{max}, for which inequality $Q_E(v_{max}) < 0$ holds, we first determine the roots of $Q_E(v_{max}) = 0$. In doing so, we develop the auxiliary equation $Q_E'(v_{max}) = 0$ by replacing $v_{max}' = v_{max}^2$. We note that as the discriminant of equation $Q_E'(v_{max}) = 0$ is equal to $72c_1c_4/\Delta d^2$, which is greater than zero for Δd, $Q_E'(v_{max}) = 0$ has two distinct roots, one of which is positive and the other one is negative. Clearly, as $v_{max} = \pm\sqrt{v_{max}'}$, the negative root does not give a valid real value for v_{max}, whereas the positive one does. Therefore, $Q_E(v_{max})$ has only one positive root. Note that $Q_E(v_{max}) < 0$ holds for $v_{max} < v_2'$, where v_2' is the positive root of $Q_E(v_{max}) = 0$. The reason for this is that $Q_E(0) < 0$ and $\frac{\partial^2 Q_E(v_{max})}{\partial v_{max}^2} > 0$, thus implying that $Q_E(0)$ is negative for $0 < v_{max} < v_2'$. Therefore, inequality (A.9) holds if

$$v_{max} < \overbrace{\sqrt{\frac{-c_2 + \sqrt{c_2^2 - 4\left(\dfrac{6c_1}{\Delta d^2}\right)(-3c_4)}}{2\left(\dfrac{6c_1}{\Delta d^2}\right)}}}^{v_2'}. \tag{A.10}$$

Subsequently, in order to satisfy Eq. (A.8), we have

$$v_{max} < \overbrace{\min\{v_1', v_2'\}}^{v_1}, \tag{A.11}$$

for which the right-hand is equal to v_2' because it is straightforward to show that $v_2' < v_1'$ for $\Delta d > 0$ given the experiment-driven values of $\{c_i\}_{i=1}^6$ taken from Tokekar et al. (2014). For illustration, $\mathbf{A_1} \in \mathbb{R}_+^2$ depicts the region that satisfies inequality (A.11), thus representing the values of $(\Delta d, v_{max})$, for which $E(\omega_d)$ has a local minimum $\forall \omega_d \in (0,1)$ (see Fig. 3.4). $\qquad \square$

A.3 Proof of Lemma 3.3

Proof: $g(\omega_d) = \frac{\partial f(\omega_d)}{\partial \omega_d}$ is a continuous function of ω_d^*, thereby having a root in interval $(0,1)$ if and only if $g(0)g(1) < 0$, which implies that

$$(1 - K')K' < 0 \Rightarrow 0 < K' < 1. \tag{A.12}$$

The left-hand inequality, $0 < K'$, reduces to

$$\overbrace{A_1 v_{max}^4 + B_1 v_{max}^2 + C_1 v_{max} + D_1}^{Q_1(v_{max})} > 0, \tag{A.13}$$

where $A_1, B_1, C_1,$ and D_1 are given in Eq. (3.29). Note that in order to evaluate the range of v_{max} for which inequality (A.12) holds, we have to determine the location of the roots (i.e., zeros) of equation $Q_1(v_{max}) = 0$. We also note that $Q_1(v_{max}) = 0$ is a quartic equation that has four roots, two of which are real while the other two are complex, as its discriminant is negative. Further, as $Q_1(0) > 0$ holds and $\lim_{v_{max} \to +\infty} Q_1(v_{max}) = -\infty$, we conclude that one of the real roots is positive while the other one is negative. Therefore, inequality (A.12) holds for $v_{max} < z_m$, where z_m is the (only) positive root of $Q_1(v_{max}) = 0$.

Next, we turn our attention to the right-hand side inequality, $K' < 1$, which reduces to

$$2c_1 v_{max}^3 < \Delta d \left(\frac{\Delta d c_4}{v_{max}} - \frac{\Delta d v_{max} c_2}{3} + \frac{E_m}{2} \right)$$

$$\stackrel{v_{max}>0}{\Leftrightarrow} \overbrace{A_2 v_{max}^4 + B_2 v_{max}^2 + C_2 v_{max} + D_2}^{Q_2(v_{max})} > 0, \tag{A.14}$$

where

$$A_2 = A_1 \Delta d - 2c_1,$$
$$B_2 = B_1 \Delta d,$$
$$C_2 = C_1 \Delta d,$$
$$D_2 = D_1 \Delta d. \tag{A.15}$$

We note that $Q_2(v_{max})$ is greater than zero only for $v_{max} < z'_m$, where z'_m is the (only) positive root of $Q_2(v_{max}) = 0$. The reason for this is that as $Q_2(v_{max}) = 0$ has two real roots, one of which is positive and the other one is negative, and $Q_2(0) > 0$, $Q_2(v_{max})$ is greater than zero for $v_{max} < z'_m$. Clearly, in order for both right- and left-hand inequalities in (A.12) to hold, v_{max} has to be smaller than min $\{z_m, z'_m\}$. We note that for $\Delta d > 0$ we have $z'_m < z_m$, therefore min $\{z_m, z'_m\} = z'_m$. Standing as the only positive root of $Q_2(v_{max})$, z'_m is max $_{Z'_i>0:\,\Im m[Z'_i]=0}\{Z'_i\}$. Then, $g(\omega_d) = 0$ has a root ω_d^* in interval $(0,1)$ if and only if $v_{max} < $ max $_{Z'_i>0:\,\Im m[Z'_i]=0}\{Z'_i\}$, $\forall \Delta d > 0$. $\qquad\square$

A.4 Proof of Lemma 5.1

Proof: To compute the average channel access delay, we define a two-dimensional Markov process $(s(t), b(t))$ shown in Fig. A.1 under unsaturated conditions and estimate the average service time Δ_i of MU i in a WLAN using the IEEE 802.11 distributed coordination function (DCF) for access control, whereby $b(t)$ and $s(t)$ denote the random back-off counter and size of the contention window at time t, respectively. Without loss of generality, let us focus on a tagged user and drop the subscript i for now. Let P_f and W_s denote the probability of a failed transmission attempt (i.e., collision or erroneous transmission) and contention window size at back-off stage s, respectively. Note that the back-off stage s is incremented after each failed attempt up to the maximum value m, while the contention window is doubled at each stage, i.e., $W_s = 2^s W_0$.

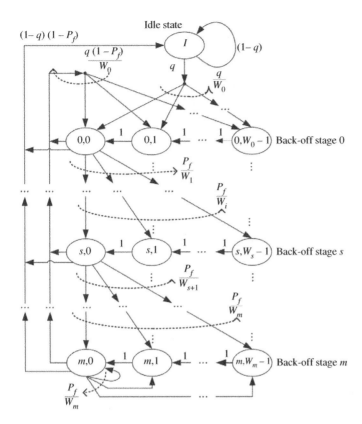

Figure A.1 Two-dimensional Markov chain for distributed coordination function (DCF) contention model under unsaturated traffic conditions. Source: Ebrahimzadeh and Maier (2020). © 2020 IEEE.

From the viewpoint of a WiFi user, collisions may occur with probability p_c on transmitted packets, while erroneous transmission attempts may happen with probability p_e. Assuming that the collided and erroneous transmission events are statistically independent, a packet is successfully transmitted after a collision-free attempt followed by an error-free transmission. The probability of a successful transmission is therefore equal to $(1 - p_e)(1 - p_c)$, from which we infer that the probability P_f of a failed transmission attempt is computed as follows:

$$P_f = 1 - \left(1 - p_c\right)\left(1 - p_e\right). \tag{A.16}$$

A WiFi user is in idle state, if (*i*) a successfully transmitted packet leaves the system without any waiting packet in the queue and (*ii*) no packet arrives during the current time slot given that the user was in idle state in the preceding time slot. We note that for Poisson arrival these two events are identical and equal to $1 - q$.

With these considerations, we move on to analyze the Markov model in Fig. A.1, where $m + 1$ different back-off stages are considered. The maximum contention window size is $2^m W_0$. Transmissions are attempted only in $(s, 0)$ states $(s = 0, 1, ..., m)$. Upon a failed transmission attempt in state $(s, 0)$, there will be a transition to new state $(s + 1, k)$, where k is uniformly selected from $[0, W_{s+1}]$. From state $(s, 0)$, we enter the initial back-off stage, again given that the transmission is successful and the buffer is still nonempty; otherwise, we transit in the idle state I and wait for an incoming packet.

The transition probabilities of the two-dimensional Markov chain in Fig. A.1 are computed as follows:

$$P_{(s,k)|(s,k+1)} = 1; \qquad \forall k \in [0, W_s - 2], s \in [0, m] \tag{A.17a}$$

$$P_{(0,k)|(s,0)} = \frac{q(1 - P_f)}{W_0}; \qquad \forall k \in [0, W_0 - 1], s \in [0, m] \tag{A.17b}$$

$$P_{(s,k)|(s-1,0)} = \frac{P_f}{W_s}; \qquad \forall k \in [0, W_0 - 1], s \in [1, m] \tag{A.17c}$$

$$P_{(m,k)|(m,0)} = \frac{P_f}{W_m}; \qquad \forall k \in [0, W_m - 1] \tag{A.17d}$$

$$P_{I|(s,0)} = (1 - q)(1 - P_f); \qquad \forall s \in [0, m] \tag{A.17e}$$

$$P_{(0,k)|I} = \frac{q}{W_0}; \qquad \forall k \in [0, W_0 - 1] \tag{A.17f}$$

$$P_{I|I} = 1 - q, \tag{A.17g}$$

where $P_{(a,b)|(c,d)}$ denotes the transition probability from state $(s(t) = c, b(t) = d)$ at time t to state $(s(t + 1) = a, b(t + 1) = b)$ at time $t + 1$.

In order to find the stationary distributions

$$b_{s,k} = \lim_{k \to \infty} P(s(t) = s, b(t) = k), \forall k \in [0, W_s - 1], s \in [0, m],$$

we consider Eqs. (A.17) together with the normalization equation

$$b_I + \sum_s \sum_k b_{s,k} = 1,$$

where b_I denote the stationary probability that the WiFi user is in idle state. After finding the stationary distributions, the probability τ that a WiFi user attempts to transmit in a given time slot is then obtained as

$$\tau = \sum_{s=0}^{m} b_{s,0} = \frac{1}{\frac{W_0+1}{2} + \frac{W_0 P_f \left(1 - (2P_f)^m\right)}{2(1 - 2P_f)q} + \frac{(1-q)(1-P_f)}{q}}. \tag{A.18}$$

From a system point of view, WiFi subscriber i does not experience a collision if the remaining users do not attempt to transmit, thus $1 - p_{c_i} = \prod_{v:v \neq i} (1 - \tau_v)$. Moreover, $p_{e,i}$ is estimated by $1 - (1 - p_b)^{\overline{L}_i}$, where $\overline{L}_i$ and p_b is the average length of a packet transmitted by WiFi user i and bit error probability, respectively.

The probability of a collision-free packet transmission P_s provided that there is at least one transmission attempt is given by $\frac{1}{P_{tr}} \left(\sum_i \tau_i \prod_{v,v \neq i} (1 - \tau_v) \right)$, whereby the probability P_{tr} that there is at least one transmission attempt is equal to $1 - \prod_i (1 - \tau_i)$. The average slot duration E_s is then obtained as

$$E_s = (1 - P_{tr}) \epsilon + P_{tr} (1 - P_s) T_c + P_{tr} P_s P_e T_e + P_{tr} P_s (1 - P_e) T_s, \qquad (A.19)$$

where T_c, T_e, and T_s are given in Aurzada et al. (2014). We also note that q can be approximated as follows

$$q = 1 - e^{-\lambda E_s}, \qquad (A.20)$$

whereby E_s is given in Eq. (A.19). In order to obtain the steady-state values of q, P_f, τ, and E_s, and Δ_i, we numerically solve the system of non-linear equations (5.15), (A.18), (A.19), and (A.20). □

Bibliography

ABI Research and Qualcomm (2017). Augmented and virtual reality: the first wave of 5G killer apps. White Paper.

Ahmad, I., Kumar, T., Liyanage, M. et al. (2018). Towards gadget-free internet services: a roadmap of the naked world. *Elsevier Telematics and Informatics* 35 (1): 82–92.

Aijaz, A., Dohler, M., Aghvami, A.H. et al. (2017). Realizing the Tactile Internet: haptic communications over next generation 5G cellular networks. *IEEE Wireless Communications* 24 (2): 82–89.

Aijaz, A., Dawy, Z., Pappas, N. et al. (2018). Toward a Tactile Internet Reference Architecture: Vision and Progress of the IEEE P1918.1 Standard. *arXiv preprint arXiv:1807.11915*.

Andrews, J.G. (2013). Seven ways that HetNets are a cellular paradigm shift. *IEEE Communications Magazine* 51 (3): 136–144.

Andrews, J.G., Buzzi, S., Choi, W. et al. (2014). What will 5G be? *IEEE Journal on Selected Areas in Communications* 32 (6): 1065–1082.

Antonakoglou, K., Xu, X., Steinbach, E. et al. (2018). Toward haptic communications over the 5G Tactile Internet. *IEEE Communications Surveys & Tutorials* 20 (4): 3034–3059.

Audi and VGo (2014). http://www.vgocom.com/audi (accessed 05 April 2020).

Aung, Y.N. and Tantidham, T. (2017). Review of ethereum: smart home case study. *Proceedings of IEEE International Conference on Information Technology (INCIT)*, pp. 1–4.

Aurzada, F., Lévesque, M., Maier, M., and Reisslein, M. (2014). FiWi access networks based on next-generation PON and gigabit-class WLAN technologies: a capacity and delay analysis. *IEEE/ACM Transactions on Networking* 22 (4): 1176–1189.

Barrientos, A.G., Lopez, J.L., Espinoza, E.S. et al. (2016). Object transportation using a cooperative mobile multi-robot system. *IEEE Latin America Transactions* 14 (3): 1184–1191.

Toward 6G: A New Era of Convergence, First Edition. Amin Ebrahimzadeh and Martin Maier.
© 2021 The Institute of Electrical and Electronics Engineers, Inc.
Published 2021 by John Wiley & Sons, Inc.

Basar, E., Di Renzo, M., De Rosny, J. et al. (2019). Wireless communications through reconfigurable intelligent surfaces. *IEEE Access* 7: 116753–116773.

Bastug, E., Bennis, M., Médard, M., and Debbah, M. (2017). Toward interconnected virtual reality: opportunities, challenges, and enablers. *IEEE Communications Magazine* 55 (6): 110–117.

Beck, R. (2018). Beyond bitcoin: the rise of blockchain world. *IEEE Computer* 51 (2): 54–58.

Beyranvand, H., Lévesque, M., Maier, M. et al. (2017). Toward 5G: FiWi enhanced LTE-A HetNets with reliable low-latency fiber backhaul sharing and WiFi offloading. *IEEE/ACM Transactions on Networking* 25 (2): 690–707.

Bi, Q. (2019). Ten trends in the cellular industry and an outlook on 6G. *IEEE Communications Magazine* 57 (12): 31–36.

Biermann, T., Scalia, L., Choi, C. et al. (2013). How backhaul networks influence the feasibility of coordinated multipoint in cellular networks. *IEEE Communications Magazine* 51 (8): 168–176.

Bohm, D. (2002). *Wholeness and the Implicate Order*. Routledge.

Bradshaw, J.M., Dignum, V., Jonker, C.M., and Sierhuis, M. (2012). Human-agent-robot teamwork. *Proceedings of ACM/IEEE International Conference on Human-Robot Interaction (HRI)*, p. 487.

Brucker, P. (2007). *Scheduling Algorithms*. Springer.

Brynjolfsson, E. and McAfee, A. (2014). *The Second Machine Age: Work, Progress, and Prosperity in a Time of Brilliant Technologies*. W. W. Norton & Company.

Buonomano, D. (2017). *Your Brain is a Time Machine: The Neuroscience and Physics of Time*. W. W. Norton.

Buterin, V. (2013). A next-generation smart contract & decentralized application platform. Ethereum White Paper. www.ethereum.org (accessed 5 March 2020).

Buzacott, J.A. (1996). Commonalities in reengineered business processes: models and issues. *Management Science* 42 (5): 768–782.

Cabrera, J.A., Schmoll, R., Nguyen, G.T. et al. (2019). Softwarization and network coding in the mobile edge cloud for the Tactile Internet. *Proceedings of the IEEE* 107 (2): 350–363.

Cacciapuoti, A.S., Caleffi, M., Van Meter, R., and Hanzo, L. (2020a). When[SJ2] entanglement meets classical communications: quantum teleportation for the quantum internet (invited paper). *IEEE Transactions on Communications* 68 (6).

Cacciapuoti, A.S., Caleffi, M., Tafuri, F. et al. (2020). Quantum internet: networking challenges in distributed quantum computing. *IEEE Network* 34 (1): 137–143.

Chang, C., Wang, S., and Wang, C. (2016). Exploiting moving objects: multi-robot simultaneous localization and tracking. *IEEE Transactions on Automation Science and Engineering* 13 (2): 810–827.

Chen, X., Jiao, L., Li, W., and Fu, X. (2016). Efficient multi-user computation offloading for mobile-edge cloud computing. *IEEE/ACM Transactions on Networking* 24 (5): 2795–2808.

Chen, M.H., Liang, B., and Dong, M. (2017). Joint offloading and resource allocation for computation and communication in mobile cloud with computing access point. *Proceedings of IEEE INFOCOM*, pp. 1–9.

Chen, Z., Ma, X., Zhang, B. et al. (2019). A survey on terahertz communications. *China Communications* 16 (2): 1–35.

Chowdhury, M., Steinbach, E., Kellerer, W., and Maier, M. (2018). Context-aware task migration for HART-centric collaboration over FiWi based Tactile Internet infrastructures. *IEEE Transactions on Parallel and Distributed Systems* 29 (6): 1231–1246.

Daugherty, P.R. and Wilson, H.J. (2018). *Human + Machine: Reimagining Work in the Age of AI*. Harvard Business Review Press.

Davenport, T.H. and Kirby, J. (2016). *Only Humans Need Apply: Winners and Losers in the Age of Smart Machines*. Harper Business.

David, K. and Berndt, H. (2018). 6G vision and requirements: is there any need for beyond 5G? *IEEE Vehicular Technology Magazine* 13 (3): 72–80.

David, K., Elmirghani, J., Haas, H., and You, X. (2019). Defining 6G: challenges and opportunities. *IEEE Vehicular Technology Magazine* 14 (3): 14–16.

Dohler, M., Mahmoodi, T., Lema, M.A. et al. (2017). Internet of skills, where robotics meets AI, 5G and the Tactile Internet. *Proceedings of European Conference on Networks and Communications (EuCNC)*, pp. 1–5.

Duan, X., Yan, Z., Geng, G., and Yan, B. (2018). DNSLedger: decentralized and distributed name resolution for ubiquitous IoT. *Proceedings of IEEE International Conference on Consumer Electronics (ICCE)*, pp. 1–3.

Dublon, G. and Paradiso, J.A. (2014). Extra sensory perception. *Scientific American* 311 (1): 36–41.

Ebrahimzadeh, A. and Maier, M. (2019). Delay-constrained teleoperation task scheduling and assignment for human+machine hybrid activities over FiWi enhanced networks. *IEEE Transactions on Network and Service Management* 16 (4): 1840–1854.

Ebrahimzadeh, A. and Maier, M. (2020). Cooperative computation offloading in FiWi enhanced 4G HetNets using self-organizing MEC. *IEEE Transactions on Wireless Communications* 19 (7): 4480–4493.

Elhajj, I., Xi, N., Fung, W.K. et al. (2001). Haptic information in internet-based teleoperation. *IEEE/ASME Transactions on Mechatronics* 6 (3): 295–304.

Elizondo Leal, J.C., Ramirez Torres, J.G., Rodriguez Tello, E., and Martinez Angulo, J.R. (2016). Multi-robot exploration using self-biddings under constraints on communication range. *IEEE Latin America Transactions* 14 (2): 971–982.

Engelbart, D.C. (1962). Augmenting Human Intellect: A Conceptual Framework. *Stanford Research Institute Summary Report AFOSR-3233.*

Fan, Q. and Ansari, N. (2018). Workload allocation in hierarchical cloudlet networks. *IEEE Communications Letters* 22 (4): 820–823.

Fast-Berglund, Å., Gong, L., and Li, D. (2018). Testing and validating extended reality (xR) technologies in manufacturing. *Procedia Manufacturing* 25: 31–38.

Fernández-Caramés, T.M. and Fraga-Lamas, P. (2018). A review on the use of blockchain for the internet of things. *IEEE Access* 6: 32979–33001.

Fettweis, G.P. (2014). The Tactile Internet: applications and challenges. *IEEE Vehicular Technology Magazine* 9 (1): 64–70.

Fettweis, G. and Alamouti, S. (2014). 5G: personal mobile internet beyond what cellular did to telephony. *IEEE Communications Magazine* 52 (2): 140–145.

Freeman, R.B. (2016). Who owns the robots rules the world. *Harvard Magazine* 118 (5): 37–39.

Ghazisaidi, N. and Maier, M. (2011). Hierarchical frame aggregation techniques for hybrid fiber-wireless access networks. *IEEE Communications Magazine* 49 (9): 64–73.

Giordani, M. and Zorzi, M. (2020). Satellite communication at millimeter waves: a key enabler of the 6G era. *Proceedings of IEEE International Conference on Computing, Networking and Communications (ICNC)*, pp. 383–388.

Green, P. (2001). Progress in optical networking. *IEEE Communications Magazine* 39 (1): 54–61.

Gupta, Y., Shorey, R., Kulkarni, D., and Tew, J. (2018). The applicability of blockchain in the internet of things. *Proc. IEEE International Conference on Communication Systems Networks (COMSNETS)*, pp. 561–564.

Guo, H. and Liu, J. (2018). Collaborative computation offloading for multiaccess edge computing over fiber-wireless networks. *IEEE Transactions on Vehicular Technology* 67 (5): 4514–4526.

Guo, S., Xiao, B., Yang, Y., and Yang, Y. (2016). Energy-efficient dynamic offloading and resource scheduling in mobile cloud computing. *Proceedings of IEEE INFOCOM*, pp. 1–9.

Haddadin, S., Johannsmeier, L., and Díaz Ledezma, F. (2019). Tactile robots as a central embodiment of the Tactile Internet. *Proceedings of the IEEE* 107 (2): 471–487.

Han, Y.S., Deng, J., and Haas, Z.J. (2006). Analyzing multi-channel medium access control schemes with ALOHA reservation. *IEEE Transactions on Wireless Communications* 5 (8): 2143–2152.

Hornik, K., Stinchcombe, M., and White, H. (1989). Multilayer feedforward networks are universal approximators. *Neural Networks* 2 (5): 359–366.

Hossain, A.D., Ummy, M., Hossain, A., and Kouar, M. (2017). Revisiting FiWi: on the merits of a distributed upstream resource allocation scheme. *IEEE/OSA Journal of Optical Communications and Networking* 9 (9): 773–781.

Huang, T., Yang, W., Wu, J. et al. (2019). A survey on green 6G network: architecture and technologies. *IEEE Access* 7: 175758–175768.

ITU-T Supplement G.Sup.5GP (2018). 5G Wireless Fronthaul Requirements in a PON Context.

ITU-T Technology Watch Report (2014). The Tactile Internet.

Kafaie, S., Ahmed, M.H., Chen, Y., and Dobre, O.A. (2018). Performance analysis of network coding with IEEE 802.11 DCF in multi-hop wireless networks. *IEEE Transactions on Mobile Computing* 17 (5): 1148–1161.

Kani, J., Bourgart, F., Cui, A. et al. (2009). Next-generation PON-Part I: Technology roadmap and general requirements. *IEEE Communications Magazine* 47 (11): 43–49.

Kasparov, G. (2017). *Deep Thinking: Where Machine Intelligence Ends and Human Creativity Begins*. Public Affairs.

Kato, N., Mao, B., Tang, F. et al. (2020). Ten challenges in advancing machine learning technologies toward 6G. *IEEE Wireless Communications* 27 (3): 96–103.

Katz, M., Matinmikko-Blue, M., and Latva-Aho, M. (2018). 6 Genesis flagship program: building the bridges towards 6G-enabled wireless smart society and ecosystem. *Proceedings of IEEE 10th Latin-American Conference on Communications (LATINCOM)*, 1–9.

Kehoe, B., Patil, S., Abbeel, P., and Goldberg, K. (2015). A survey of research on cloud robotics and automation. *IEEE Transactions on Automation Science and Engineering* 12 (2): 398–409.

Kelly, K. (2016). *The Inevitable: Understanding the 12 Technological Forces that Will Shape Our Future*. Viking.

Kim, H. (2018). RoF-based optical fronthaul technology for 5G and beyond. *Proceedings of Optical Fiber Communications Conference and Exposition (OFC)*, pp. 1–3.

Klaine, P.V., Imran, M.A., Onireti, O., and Souza, R.D. (2017). A survey of machine learning techniques applied to self-organizing cellular networks. *IEEE Communications Surveys & Tutorials* 19 (4): 2392–2431.

Latva-aho, M. and Leppänen, K. (eds.) (2019). 6G White Paper, Key Drivers and Research Challenges for 6G Ubiquitous Wireless Intelligence.

Lee, D., Zaheer, S.A., and Kim, J. (2014). Ad Hoc network-based task allocation with resource-aware cost generation for multi-robot systems. *IEEE Transactions on Industrial Electronics* 61 (12): 6871–6881.

Lema, M.A., Laya, A., Mahmoodi, T. et al. (2017). Business case and technology analysis for 5G low latency applications. *IEEE Access* 5: 5917–5935.

Leontief, W. (1983). Technological advance, economic growth, and the distribution of income. *Population and Development Review* 9 (3): 403–410.

Letaief, K.B., Chen, W., Shi, Y. et al. (2019). The roadmap to 6G: AI empowered wireless networks. *IEEE Communications Magazine* 57 (8): 84–90.

Li, J. and Chen, J. (2017). Passive optical network based mobile backhaul enabling ultra-low latency for communications among base stations. *IEEE/OSA Journal of Optical Communications and Networking* 9 (10): 855–863.

Liao, C., Bao, S., Cheng, C., and Chen, K. (2017). On design issues and architectural styles for blockchain-driven IoT services. *Proceedings of IEEE International Conference on Consumer Electronics - Taiwan (ICCE-TW)*, pp. 351–352.

Licklider, J.C.R. (1960). Man-computer symbiosis. *IRE Transactions on Human Factors in Electronics* HFE-1: 4–11.

Lindley, D.V. (1952). The theory of queues with a single server. *Mathematical Proceedings of the Cambridge Philosophical Society* 48 (2): 277–289.

Liu, Y., Liu, M., and Deng, J. (2013). Evaluating opportunistic multi-channel MAC: is diversity gain worth the pain? *IEEE Journal on Selected Areas in Communications* 31 (11): 2301–2311.

Liu, J., Guo, H., Nishiyama, H. et al. (2016). New perspectives on future smart FiWi networks: scalability, reliability, and energy efficiency. *IEEE Communications Surveys & Tutorials* 18 (2): 1045–1072.

Liu, L., Chang, Z., Guo, X. et al. (2018). Multiobjective optimization for computation offloading in fog computing. *IEEE Internet of Things Journal* 5 (1): 283–294.

Luo, R.C., Lee, W.Z., Chou, J.H., and Leong, H.T. (1999). Telecontrol of rapid prototyping machine via internet for automated telemanufacturing. *Proceedings of IEEE International Conference on Robotics and Automation*, pp. 2203–2208.

Maier, M. (2014). The escape of sisyphus or what "post NG-PON2" should do apart from neverending capacity upgrades. *Photonics, Special Issue of All Optical Networks for Communications* 1 (1): 47–66.

Maier, M. and Ebrahimzadeh, A. (2019). Towards immersive Tactile Internet experiences: low-latency FiWi enhanced mobile networks with edge intelligence [Invited]. *IEEE/OSA Journal of Optical Communications and Networking, Special Issue on Latency in Edge Optical Networks* 11 (4): B10–B25.

Maier, M. and Ghazisaidi, N. (2018). *FiWi Access Networks*. Cambridge: Cambridge University Press.

Maier, M. and Rimal, B.P. (2015). The audacity of fiber-wireless (FiWi) networks: revisited for clouds and cloudlets (Invited Paper). *China Communications* 12 (8): 33–45.

Maier, M., Ghazisaidi, N., and Reisslein, M. (2008). The audacity of fiber-wireless (FiWi) networks (Invited Paper). In *Proceedings of ICST International Conference on Access Networks (AccessNets)*, pp. 1–10.

Maier, M., Chowdhury, M., Rimal, B.P., and Van, D.P. (2016). The Tactile Internet: vision, recent progress, and open challenges. *IEEE Communications Magazine* 54 (5): 138–145.

Maier, M., Ebrahimzadeh, A., and Chowdhury, M. (2018). The Tactile Internet: automation or augmentation of the human? *IEEE Access* 6: 41607–41618.

McAfee, A. and Brynjolfsson, E. (2017). *Machine, Platform, Crowd: Harnessing Our Digital Future*. New York, NY: W. W. Norton.

McGuire, S., Furlong, P.M., Heckman, C. et al. (2018). Failure is not an option: policy learning for adaptive recovery in space operations. *IEEE Robotics and Automation Letters* 3 (3): 1639–1646.

McKinsey Global Institute (2017). *A Future that Works: Automation, Employment, and Productivity*. McKinsey Global Institute Research, Tech. Rep 60.

Medepalli, K. and Tobagi, F.A. (2006). Towards performance modeling of IEEE 802.11 based wireless networks: a unified framework and its applications. *Proceedings of IEEE INFOCOM*, pp. 1–12.

Meli, L., Pacchierotti, C., and Prattichizzo, D. (2017). Experimental evaluation of magnified haptic feedback for robot-assisted needle insertion and palpation. *The International Journal of Medical Robotics and Computer Assisted Surgery* 13 (4): e1809.

Miettinen, A.P. and Nurminen, J.K. (2010). Energy efficiency of mobile clients in cloud computing. Proceedings of the 2nd USENIX Conference on Hot Topics in Cloud Computing. Boston, MA, Berkeley, CA, USA, pp. 1–7.

Nissan Newsroom (2019). Nissan unveils invisible-to-visible technology concept at CES: future connected-car technology merges real and virtual worlds to help drivers 'see the invisible'. (accessed 05 April 2020).

Novo, O. (2018). Blockchain meets IoT: an architecture for scalable access management in IoT. *IEEE Internet of Things Journal* 5 (2): 1184–1195.

Nowak, P. (2017). Nissan uses NASA rover tech to remotely oversee autonomous car. *New Scientist* (accessed 05 April 2020).

Özyılmaz, K.R. and Yurdakul, A. (2017). The applicability of blockchain in the internet of things. *Proceedings of IEEE International Conference on Communication Systems Networks (COMSNETS)*, pp. 1–2.

Pan, J., Wang, J., Hester, A. et al. (2019). EdgeChain: an edge-IoT framework and prototype based on blockchain and smart contracts. *IEEE Internet of Things Journal* 6 (3): 4719–4732.

Pérez, G.O., Hernández, J.A., and Larrabeiti, D. (2018). Fronthaul network modeling and dimensioning meeting ultra-low latency requirements for 5G. *IEEE/OSA Journal of Optical Communications and Networking* 10 (6): 573–581.

Pfeiffer, T. (2018). Can PON technologies accelerate 5G deployments? *Proceedings of IEEE Optical Network Design and Modelling (ONDM), Workshop on Optical Technologies in the 5G Era*.

Pham, P.P., Perreau, S., and Jayasuriya, A. (2005). New cross-layer design approach to Ad Hoc networks under Rayleigh fading. *IEEE Journal on Selected Areas in Communications* 23 (1): 28–39.

Pine, B.J. II and Korn, K.C. (2011). *Infinite Possibility: Creating Customer Value on the Digital Frontier*. Berrett-Koehler Publishers.

Polyzos, G.C. and Fotiou, N. (2017). Blockchain-assisted information distribution for the internet of things. *Proceedings of IEEE International Conference on Information Reuse and Integration (IRI)*, pp. 75–78.

Prattichizzo, D., Shinoda, H., Tan, H.Z. et al. (2018). *Haptics: Science, Technology, and Applications, Lecture Notes in Computer Science (LNCS)*, vol. 10894, 4–11. Springer.

Ranaweera, C., Resende, M.G.C., Reichmann, K. et al. (2013a). Design and optimization of fiber optic small-cell backhaul based on an existing fiber-to-the-node residential access network. *IEEE Communications Magazine* 51 (9): 62–69.

Ranaweera, C.S., Iannone, P.P., Oikonomou, K.N. et al. (2013b). Design of cost-optimal passive optical networks for small cell backhaul using installed fibers [Invited]. *IEEE/OSA Journal of Optical Communications and Networking* 5 (10): A230–A239.

Rappaport, T.S., Xing, Y., Kanhere, O et al. (2019). Wireless communications and applications above 100 GHz: opportunities and challenges for 6G and beyond (invited paper). *IEEE Access* 7: 78729–78757.

Rimal, B.P., Van, D.P., and Maier, M. (2017a). Cloudlet enhanced fiber-wireless access networks for mobile-edge computing. *IEEE Transactions on Wireless Communications* 16 (6): 3601–3618.

Rimal, B.P., Van, D.P., and Maier, M. (2017b). Mobile-edge computing versus centralized cloud computing over a converged FiWi access network. *IEEE Transactions on Network and Service Management* 14 (3): 498–513.

Rimal, B.P., Van, D.P., and Maier, M. (2017c). Mobile edge computing empowered fiber-wireless access networks in the 5G era. *IEEE Communications Magazine* 55 (2): 192–200.

Rimal, B.P., Maier, M., and Satyanarayanan, M. (2018). Experimental testbed for edge computing in fiber-wireless broadband access networks. *IEEE Communications Magazine* 56 (8): 160–167.

Rodrigues, T.G., Suto, K., Nishiyama, H., and Kato, N. (2017). Hybrid method for minimizing service delay in edge cloud computing through VM migration and transmission power control. *IEEE Transactions on Computers* 66 (5): 810–819.

Rodrigues, T.G., Suto, K., Nishiyama, H. et al. (2018). Cloudlets activation scheme for scalable mobile edge computing with transmission power control and virtual machine migration. *IEEE Transactions on Computers* 67 (9): 1287–1300.

Ross, S.M. (2014). *Introduction to Probability Models*. Academic Press.

Rowell, C. and Han, S. (2015). Practical large scale antenna systems for 5G cellular networks. *Proceedings of IEEE International Wireless Symposium (IWS)*, pp. 1–4.

Saad, W., Bennis, M., and Chen, M. (2020). A vision of 6G wireless systems: applications, trends, technologies, and open research problems. *IEEE Network* 34 (3): 134–142.

Salman, T., Zolanvari, M., Erbad, A. et al. (2019). Security services using blockchains: a state of the art survey. *IEEE Communications Surveys & Tutorials* 21 (1): 858–880.

Satyanarayanan, M. (2017). The emergence of edge computing. *IEEE Computer* 50 (1): 30–39.

Simsek, M., Aijaz, A., Dohler, M. et al. (2016). 5G-enabled Tactile Internet. *IEEE Journal on Selected Areas in Communications* 34 (3): 460–473.

Steinbach, E., Hirche, S., Ernst, M. et al. (2012). Haptic communications. *Proceedings of the IEEE* 100 (4): 937–956.

Strinati, E.C., Barbarossa, S., Gonzalez-Jimenez, J.L. et al. (2019). 6G: the next frontier: from holographic messaging to artificial intelligence using subterahertz and visible light communication. *IEEE Vehicular Technology Magazine* 14 (3): 42–50.

Sun, X. and Ansari, N. (2017). Latency aware workload offloading in the cloudlet network. *IEEE Communications Letters* 21 (7): 1481–1484.

Taleb, T., Samdanis, K., Mada, B. et al. (2017). On multi-access edge computing: a survey of the emerging 5G network edge cloud architecture and orchestration. *IEEE Communications Surveys & Tutorials* 19 (3): 1657–1681.

Tan, H., Han, Z., Li, X.Y., and Lau, F.C.M. (2017). Online job dispatching and scheduling in edge-clouds. *Proceedings of IEEE INFOCOM*, pp. 1–9.

Tang, F., Kawamoto, Y., Kato, N., and Liu, J. (2020a). Future intelligent and secure vehicular network toward 6G: machine-learning approaches (invited paper). *Proceedings of the IEEE* 108 (2): 292–307.

Tang, W., Chen, M.Z., Dai, J.Y. et al. (2020b). Wireless communications with programmable metasurface: new paradigms, opportunities, and challenges on transceiver design. *IEEE Wireless Communications* 27 (2): 180–187.

Tapscott, D. and Tapscott, A. (2016). Blockchain revolution: how the technology behind bitcoin is changing money, business, and the world. Portfolio, Toronto, ON, Canada.

Tereshchuk, V., Stewart, J., Bykov, N. et al. (2019). An efficient scheduling algorithm for multi-robot task allocation in assembling aircraft structures. *IEEE Robotics and Automation Letters* 4 (4): 3844–3851.

Thyagaturu, A.S., Mercian, A., McGarry, M.P. et al. (2016). Software defined optical networks (SDONs): a comprehensive survey. *IEEE Communications Surveys & Tutorials* 18 (4): 2738–2786.

Tokekar, P., Karnad, N., and Isler, V. (2014). Energy-optimal trajectory planning for car-like robots. *Autonomous Robots* 37 (3): 279–300.

Tong, L., Li, Y., and Gao, W. (2016). A hierarchical edge cloud architecture for mobile computing. *Proceedings of IEEE INFOCOM*, pp. 1–9.

Urien, P. (2018). Towards secure elements for trusted transactions in blockchain and blochchain IoT (BIoT) platforms (invited paper). *Proceedings of the IEEE 4th International Conference on Mobile and Secure Services (MobiSecServ)*, pp. 1–5.

Van Den Berg, D., Glans, R., De Koning, D. et al. (2017). Challenges in haptic communications over the Tactile Internet. *IEEE Access* 5: 23502–23518.

Velasco, L., Castro, A., Asensio, A. et al. (2017). Meeting the requirements to deploy cloud RAN over optical networks. *IEEE/OSA Journal of Optical Communications and Networking* 9 (3): B22–B32.

Viswanathan, H., and P. E., Mogensen (2020). Communications in the 6G Era. *IEEE Access* 8: 57063–57074.

Wang, Y., Sheng, M., Wang, X. et al. (2016). Mobile-edge computing: partial computation offloading using dynamic voltage scaling. *IEEE Transactions on Communications* 64 (10): 4268–4282.

Weber, E. (1978). *Die Lehre vom Tastsinn und Gemeingefuehl auf Versuche gegruendet.* London: Verlag Friedrich Vieweg und Sohn.

Weiser, M. (1999). ACM SIGMOBILE mobile computing and communications review. *The computer for the 21st century* 3 (3): 3–11.

Wong, E., Dias, M.P.I., and Ruan, L. (2017). Predictive resource allocation for Tactile Internet capable passive optical LANs. *IEEE/OSA Journal of Lightwave Technology* 35 (13): 2629–2641.

Xiao, Y. and Krunz, M. (2018). Distributed optimization for energy-efficient fog computing in the Tactile Internet. *IEEE Journal on Selected Areas in Communications* 36 (11): 2390–2400.

Xu, X., Schuwerk, C., Cizmeci, B., and Steinbach, E. (2016). Energy prediction for teleoperation systems that combine the time domain passivity approach with perceptual deadband-based haptic data reduction. *IEEE Transactions on Haptics* 9 (4): 560–573.

Xu, X., Zhang, X., Gao, H. et al. (2020). BeCome: blockchain-enabled computation offloading for IoT in mobile edge computing. *IEEE Transactions on Industrial Informatics* 16 (6): 4187–4195.

Yaacoub, E. and Alouini, M. (2020). A key 6G challenge and opportunity—connecting the base of the pyramid: a survey on rural connectivity (invited paper). *Proceedings of the IEEE* 108 (4): 533–582.

Yang, R., Yu, F.R., Si, P. et al. (2019). Integrated blockchain and edge computing systems: a survey, some research issues and challenges. *IEEE Communications Surveys & Tutorials* 21 (2): 1508–1532.

Yastrebova, A., Kirichek, R., Koucheryavy, Y. et al. (2018). Future networks 2030: architecture requirements. *Proceedings of IEEE International Congress on Ultra Modern Telecommunications and Control Systems and Workshops (ICUMT)*, pp. 1–8.

You, C., Huang, K., Chae, H., and Kim, B.H. (2017). Energy-efficient resource allocation for mobile-edge computation offloading. *IEEE Transactions on Wireless Communications* 16 (3): 1397–1411.

Zhang, Y., Kasahara, S., Shen, Y. et al. (2019a). Smart contract-based access control for the internet of things. *IEEE Internet of Things Journal* 6 (2): 1594–1605.

Zhang, Z., Xiao, Y., Ma, Z. et al. (2019b). 6G wireless networks: vision, requirements, architecture, and key technologies. *IEEE Vehicular Technology Magazine* 14 (3): 28–41.

Zhang, S., Xiang, C., and Xu, S. (2020). 6G: connecting everything by 1000 times price reduction. *IEEE Open Journal of Vehicular Technology* 1: 107–115.

Zhaofeng, M., Xiaochang, W., Jain, D.K. et al. (2020). A blockchain-based trusted data management scheme in edge computing. *IEEE Transactions on Industrial Informatics* 16 (3): 2013–2021.

Zheng, N., Liu, Z., Ren, P. et al. (2017). Hybrid-augmented intelligence: collaboration and cognition. *Springer Frontiers of Information Technology & Electronic Engineering* 18 (2): 153–179.

Zhou, S., Liu, X., Effenberger, F., and Chao, J. (2018). Low-latency high-efficiency mobile fronthaul with TDM-PON (Mobile PON). *IEEE/OSA Journal of Optical Communications and Networking* 10 (1): A20–A26.

Zhu, Y.-h., Lu, H.-c., and Leung, V.C.M. (2012). Access point buffer management for power saving in IEEE 802.11 WLANs. *IEEE Transactions on Network and Service Management* 9 (4): 473–486.

Zong, B., Fan, C., Wang, X. et al. (2019). 6G technologies: key drivers, core requirements, system architectures, and enabling technologies. *IEEE Vehicular Technology Magazine* 14 (3): 18–27.

Index

Toward 6G: A New Era of Convergence, First Edition. Amin Ebrahimzadeh and Martin Maier.
© 2021 The Institute of Electrical and Electronics Engineers, Inc.
Published 2021 by John Wiley & Sons, Inc.

THE COMSOC GUIDES TO COMMUNICATIONS TECHNOLOGIES

Nim K. Cheung, *Senior Editor*
Richard Lau, *Associate Editor*

The ComSoc Guide to Next Generation Optical Transport: SDH/SONET/OTN
Huub van Helvoort

The ComSoc Guide to Managing Telecommunications Projects
Celia Desmond

WiMAX Technology and Network Evolution
Kamran Etemad and Ming-Yee Lai

An Introduction to Network Modeling and Simulation for the Practicing Engineer
Jack Burbank, William Kasch, and Jon Ward

The ComSoc Guide to Passive Optical Networks: Enhancing the Last Mile Access
Stephen Weinstein, Yuanqiu Luo, and Ting Wang

Digital Terrestrial Television Broadcasting: Technology and System
Jian Song, Zhixing Yang, and Jun Wang

TV White Space: The First Step Towards Better Utilization of Frequency Spectrum
Ser Wah Oh, Yugang Ma, Edward Peh, and Ming-Hung Tao

Digital Services in the 21st Century: A Strategic and Business Perspective
Antonio Sanchez and Belen Carro

Printed and bound by CPI Group (UK) Ltd, Croydon, CR0 4YY
07/05/2021
03070486-0001